PRAISE FOR MICHAEL BARTON

From *"The Power of Regenerative Agriculture:"*

Well done book. I liked the chapter summaries at the end of each chapter. Chapter 6, was spot on in describing the challenges and barriers to implementing regenerative agriculture. The author has a passion for this subject and has provided very clear and concise data. The Afterword ends the book on a positive note and a view of hope that we'll get there.

— SRP

These guides to being environmentally responsible often come across as a little preachy. This is not one of those. It's informative and well-written, with a cheerful attitude toward ecologically friendly farming methods.

— WHIT

From *"Introduction to Soil Science:"*

This book is about a comprehensive introduction to Soil Science - you'll learn everything from the basics of soil formation and classification, to the physical, chemical, and biological properties of soil, fertility and nutrient management techniques, and of course, conservation methods. In other words, this book is your one-stop shop for all your soil-related needs. You'll learn how to properly care for your soil, how to identify its many components, and how to make sure it's in the best condition possible.

— BOGDAN IVANOV

FOSTERING A HEALTHY PLANET

FOSTERING A HEALTHY PLANET

LEARN HOW REGENERATIVE AGRICULTURE AND SOIL SCIENCE CONTRIBUTE TO A MORE RESILIENT AND SUSTAINABLE WORLD (2-IN-1 COLLECTION)

SUSTAINABLE AGRICULTURE

MICHAEL BARTON

Book
Bound Studios

This book is dedicated to all the farmers, environmentalists, and scientists working to protect and preserve our planet for future generations. Your tireless efforts and dedication to finding solutions for a healthier future are an inspiration. I appreciate your commitment to making the world a better place.

The power of nature lies in its ability to regenerate itself. We must work in harmony with the natural world to foster a healthy planet.

— UNKNOWN

CONTENTS

INTRODUCTION TO SOIL SCIENCE

INTRODUCTION

Welcome to *"Fostering a Healthy Planet,"* a 2-in-1 collection offering a wealth of agriculture and soil science knowledge. Our planet faces many environmental issues, from climate change to soil degradation, which will have severe consequences if left unaddressed for future generations. Regenerative agriculture and soil science offer a path to a healthier future by mimicking nature and promoting soil health.

The first book, *"The Power of Regenerative Agriculture,"* explores the principles and practices of regenerative agriculture, helping farmers, students, and others interested in sustainability learn how to heal and rejuvenate the soil and the surrounding ecosystem.

The second book, *"Introduction to Soil Science,"* is an in-depth look into the scientific study of soil, covering its properties, functions, and impact on plant growth and the health of the wider ecosystem.

These two books are essential resources for anyone wanting to learn more about the critical role agriculture and soil science play in maintaining the health of our planet. So, join us and start your journey to a more sustainable future!

THE POWER OF REGENERATIVE AGRICULTURE

TRANSFORMING AGRICULTURE FOR ENVIRONMENTAL, ECONOMIC, AND SOCIAL SUSTAINABILITY

INTRODUCTION

Regenerative agriculture is a growing movement gaining traction worldwide as more and more people recognize the need for a shift toward more sustainable and equitable food systems. In this chapter, we will explore the definition of regenerative agriculture, the need for these practices, and their numerous benefits. By understanding the foundations of regenerative agriculture, we can gain insights into the potential for these practices to create a more sustainable and equitable food system for the future.

Definition of Regenerative Agriculture

Regenerative agriculture is a holistic approach to farming and land management that aims to restore and enhance the health and fertility of the soil, as well as the biodiversity of the ecosystem. This approach is based on sustainability, resilience, and regenerative capacity principles. Furthermore, it seeks to create a closed-loop system in which waste is minimized, and resources are used efficiently.

In other words, regenerative agriculture aims to create a farming system that is sustainable in the long term and able to regenerate and improve the natural resources it relies upon. This includes the soil, water, and biological diversity of the ecosystem in which it is practiced. By prioritizing these principles, regenerative agriculture seeks to create a more sustainable and resilient food system that benefits the environment and the people who depend on it.

Many different practices and techniques fall under the umbrella of regenerative agriculture, including cover cropping, composting, crop rotation, intercropping, agroforestry, and managed grazing, among others. In addition, these practices are often integrated and customized to suit the specific needs and conditions of a given farm or region.

One of the key goals of regenerative agriculture is to improve the health and fertility of the soil. Soil is a vital natural resource that provides the foundation for all plant growth and is essential for producing high-quality crops. Unfortunately, many modern farming practices have degraded the soil, leading to decreased crop yields and increased reliance on synthetic inputs such as fertilizers and pesticides. Regenerative agriculture seeks to reverse this trend by using practices that enhance the soil's structure, biology, and nutrient content, such as cover cropping, composting, and reducing tillage.

In addition to improving soil health, regenerative agriculture also seeks to conserve water and reduce erosion, enhance biodiversity, and integrate livestock in a way that benefits the ecosystem's overall health. These practices can help to create a more sustainable and resilient food system that can adapt and thrive in the face of challenges such as climate change and population growth.

Overall, regenerative agriculture offers a promising approach to farming and land management that has the potential to create a more sustainable and equitable food system for the future. By incorporating these principles and practices into our farming and land management systems, we can work towards a brighter future for agriculture and the environment.

The Need for Regenerative Agriculture

The need for regenerative agriculture is pressing and urgent, as it is driven by several complex and interconnected factors that threaten our food system's long-term sustainability and resilience.

One of the primary drivers of the need for regenerative agriculture is the negative impacts of industrial agriculture on the environment. Industrial agriculture is a model of farming that relies on monoculture, synthetic inputs such as pesticides and fertilizers, and large-scale mechanization. While this model has successfully increased food production in the short term, it has also had serious negative environmental consequences. These include soil erosion, water pollution, habitat destruction, and biodiversity loss.

Another factor contributing to the need for regenerative agriculture is the declining fertility of the soil. The soil is a vital natural resource that provides the foundation for all plant growth and is essential for producing high-quality crops. Unfortunately, many modern farming practices have degraded the soil, leading to decreased crop yields and increased reliance on synthetic inputs such as fertilizers and pesticides. This trend is unsustainable and threatens our food system's long-term productivity and viability.

Finally, the negative impacts of climate change on food production are also driving the need for regenerative agriculture. Climate change is causing rising temperatures, extreme weather events, and changing rainfall patterns that seriously threaten food production. To ensure a secure and sustainable food supply in the face of these challenges, shifting towards more sustainable and regenerative practices is essential.

Overall, the need for regenerative agriculture is clear and pressing. By shifting towards more sustainable and regenerative practices, we can work towards a more secure and sustainable food system that can better withstand future challenges.

The Benefits of Regenerative Agriculture

The benefits of regenerative agriculture are extensive and far-reaching and go beyond increasing crop yields and improving food security. By focusing on the health and fertility of the soil and promoting biodiversity, regenerative agriculture can support the long-term sustainability and resilience of the food system.

One of the most significant benefits of regenerative agriculture is its ability to improve the health and fertility of the soil. Soil is a vital natural resource that provides the foundation for all plant growth and is essential for producing high-quality crops. Using practices such as cover cropping, composting, and reducing tillage, regenerative agriculture can help increase the amount of organic matter in the soil, leading to improved structure, biology, and nutrient content. This can lead to increased crop yields and improved food security, as well as support the overall health and productivity of the soil.

Another key benefit of regenerative agriculture is the reduction of synthetic inputs such as pesticides and fertilizers. These inputs can negatively impact the environment, including water pollution and the loss of biodiversity, as well as posing potential risks to human health. By promoting practices such as crop rotation and intercropping, regenerative agriculture can help reduce the need for synthetic inputs and support using natural inputs instead. This can help protect the environment and human health while supporting the food system's long-term sustainability.

In addition to its environmental benefits, regenerative agriculture also has the potential to support the conservation of biodiversity and enhance the resilience of ecosystems to the impacts of climate change. For example, practices such as polycultures, agroforestry, and livestock integration can help increase farm biodiversity and enhance habitat for pollinators and other beneficial insects. This can support the overall health and resilience of the ecosystem while contributing to the food system's long-term sustainability.

Finally, regenerative agriculture has the potential to improve the livelihoods of farmers and create stronger, more sustainable communities. By promoting holistic management and the use of managed grazing, regenerative agriculture can help improve the profitability of farms and create more sustainable and resilient communities. It also recognizes the interconnectedness of all elements of the food system. Therefore, it seeks to create a holistic approach that benefits the environment, farmers, and consumers. By shifting towards regenerative agriculture, we can create a more sustainable and equitable food system for the future.

In conclusion, regenerative agriculture is a holistic and sustainable approach to farming and land management that has the potential to create a more resilient and equitable food system. By focusing on the principles of sustainability, resilience, and regenerative capacity, regenerative agriculture seeks to restore and enhance the health and fertility of the soil, as well as the biodiversity of the ecosystem. The benefits of regenerative agriculture include improved soil health, water conservation, reduced reliance on synthetic inputs, enhanced biodiversity, and improved farmer livelihoods. However, challenges and barriers to adopting regenerative agriculture include a lack of knowledge and resources, limited market demand, regulatory barriers, and cultural and social barriers. To overcome these challenges, investing in education and training programs, research and development, policy and regulatory reform, and market development and support will be important. By working together and implementing regenerative agriculture practices, we can create a more sustainable and equitable food system for the future.

1

THE PRINCIPLES OF REGENERATIVE AGRICULTURE

The principles of regenerative agriculture are the foundation upon which these systems are built. This chapter will delve into the specific practices that make up these principles, starting with soil health and fertility and moving on to water management, biodiversity, crop rotation and intercropping, and livestock management. Understanding these principles and how they work together is essential for the success of regenerative agriculture systems, as they provide the foundation for sustainable and productive farming practices that benefit the environment, farmers, and consumers.

Soil Health and Fertility

Regenerative agriculture recognizes that soil is a living, dynamic ecosystem essential to the health and productivity of the entire food system. By focusing on soil health and fertility, regenerative agriculture seeks to create a closed-loop system in which waste is minimized, and resources are used efficiently. This includes not only the nutrients and water needed for plant growth but also the biological diversity of the soil, which plays a vital role in the ecosystem's overall health.

In addition to providing the necessary nutrients and water for plant growth, healthy soil also can sequester carbon from the atmosphere. Carbon sequestration is capturing and storing carbon dioxide from the atmosphere in the form of organic matter in the soil. This can help mitigate climate change's impacts, as carbon dioxide is a major contributor to global warming. By increasing the amount of organic matter in the soil through practices such as cover cropping and the application of compost, regenerative agriculture can help to increase soil carbon sequestration and combat climate change.

However, industrial agriculture practices often compromise soil health and fertility, which rely on monoculture, synthetic inputs such as pesticides and fertilizers, and large-scale mechanization. These practices can lead to soil erosion, nutrient depletion, and a decline in the biological diversity of the soil. By shifting towards regenerative agriculture practices that prioritize soil health and fertility, it is possible to reverse this trend and create a more sustainable and resilient food system.

In conclusion, soil health and fertility are essential for producing high-quality crops and the overall sustainability of the food system. By focusing on these principles, regenerative agriculture can create a more sustainable and resilient food system that benefits the environment and the people who depend on it.

Water Management

Water management is an important aspect of regenerative agriculture, as it seeks to conserve this vital resource while maximizing its use in crop production. You can implement several strategies to achieve these goals, including mulching, contour planting, and porous surfaces. These practices help reduce erosion and improve water use efficiency in the farm system.

Rainwater harvesting is another key strategy in water management in regenerative agriculture. By collecting and storing rainwater, farmers can have a reliable source of water for their crops, reducing the need for irrigation and other external water sources. This can be particularly important in regions with scarce water or drought conditions.

Greywater reuse is another water management strategy you can implement in regenerative agriculture. Greywater is wastewater from household activities such as washing dishes, laundry, and showers. It can be collected and treated for reuse in irrigation. This helps reduce the demand for freshwater, a limited and valuable resource. By using greywater in irrigation, farmers can save water and reduce the impact of their farming practices on the environment.

Overall, water management is a critical component of regenerative agriculture, as it helps to conserve this vital resource and improve the efficiency of its use in crop production. By implementing mulching, contour planting, rainwater harvesting, and greywater reuse strategies, farmers can create a more sustainable and resilient food system that benefits both the environment and their livelihoods.

Biodiversity

Biodiversity is the variety of different species that live within an ecosystem. It is a vital component of a healthy and functioning ecosystem. Therefore, regenerative agriculture practices prioritize the conservation of biodiversity to enhance the overall health and resilience of the ecosystem.

One way regenerative agriculture supports biodiversity is by using polycultures, which are diverse plantings that include various crop species. Polycultures can mimic the diversity of natural ecosystems and provide multiple benefits, including increased pest control, improved soil health, and increased water retention.

Agroforestry is another regenerative agriculture practice that supports biodiversity. This practice involves integrating trees and other woody plants into agricultural landscapes, creating a mosaic of different land uses that provide habitat for a wide range of species.

Integrating livestock into regenerative agriculture systems can also support biodiversity by providing a source of organic matter for the soil and improving the land's overall health and fertility. Managed grazing, in particular, can support biodiversity by mimicking the natural movements of wild herds, which can help promote the ecosystem's health.

Overall, regenerative agriculture practices that prioritize biodiversity can help create a more sustainable and resilient food system that benefits the environment and the people who depend on it.

Crop Rotation and Intercropping

Crop rotation is a key regenerative agriculture practice involving systematically planting different crops in a specific sequence on the same land. This practice has numerous benefits for soil health and fertility, as it helps improve the soil's structure, biology, and nutrient content. It can also reduce the incidence of pests and diseases, as different crops have different pest and disease pressures. In addition, crop rotation can enhance the farm's overall productivity, as it allows for a more diverse mix of crops to be grown and harvested.

Intercropping is another regenerative agriculture practice that involves planting different crops in close proximity to each other. This practice can improve soil health and fertility, as it allows for the integration of different types of plants with complementary nutrient requirements. For example, intercropping legumes with non-legumes can help fix nitrogen in the soil, which can benefit the growth of non-legume crops. Intercropping can also help to reduce pest and disease pressures, as the presence of different types of crops can create a more diverse and complex ecosystem that is less conducive to pest and disease outbreaks. In addition, intercropping can enhance the farm's overall productivity, as it allows for the simultaneous production of multiple crops on the same piece of land.

Overall, crop rotation and intercropping are key regenerative agriculture practices that can help to improve soil health and fertility, reduce pests and diseases, and enhance the farm's overall productivity. By incorporating these practices into their farming systems, farmers can work towards a more sustainable and resilient food system that benefits the environment and the people who depend on it.

Livestock Management

Regenerative agriculture systems often prioritize livestock integration in a way that benefits both the animals and the overall health of the ecosystem. You can achieve this through managed grazing, in which livestock are rotated through different pastures, allowing the grasses and other vegetation to recover and regenerate. This practice can help improve soil health and fertility, reduce erosion, and improve water retention. In addition, manure from well-managed livestock can provide a natural source of nutrients for crops, reducing the need for synthetic fertilizers.

It is important to remember that the integration of livestock into regenerative agriculture systems must be done holistically and responsibly, with the health and well-being of the animals as a top priority. This may involve the use of rotational grazing systems, the provision of clean water and high-quality feed, and the implementation of humane handling practices. Taking a holistic approach to livestock management makes it possible to create a symbiotic relationship between the animals and the ecosystem, in which both thrive.

In conclusion, regenerative agriculture offers a promising approach to farming and land management that has the potential to create a more sustainable and equitable food system for the future. By focusing on principles such as soil health and fertility, water management, biodiversity, crop rotation and intercropping, and livestock management, regenerative agriculture can create a closed-loop system in which waste is minimized, and resources are used efficiently. These practices can help improve the soil's health and fertility, conserve water and reduce erosion, enhance biodiversity, and create a more resilient and sustainable food system. In addition, regenerative agriculture has the potential to improve the livelihoods of farmers and create stronger, more sustainable communities. By incorporating these principles and practices into our farming and land management systems, we can work towards a brighter future for agriculture and the environment.

Chapter Summary

- The principles of regenerative agriculture include soil health and fertility, water management, biodiversity, crop rotation and intercropping, and livestock management.
- Soil health and fertility are essential for plant growth and carbon sequestration. Industrial agriculture practices often compromise soil health, but regenerative agriculture practices can reverse this trend.
- Water management in regenerative agriculture aims to conserve and efficiently use water through mulching, contour planting, and rainwater harvesting.
- Biodiversity is important for the overall health of the ecosystem. We can promote it through practices such as cover cropping and intercropping.
- Crop rotation and intercropping can improve soil health and fertility and reduce the need for synthetic inputs.
- Livestock management in regenerative agriculture focuses on the welfare of the animals and promoting practices such as rotational grazing.
- Regenerative agriculture can benefit the environment, farmers, and consumers by creating a closed-loop system that minimizes waste and increases resource efficiency.
- The principles of regenerative agriculture work together to create a sustainable and productive farming system.

2

THE HISTORY OF REGENERATIVE AGRICULTURE

The history of agriculture is long and complex, with the evolution of farming practices reflecting societies' changing needs and priorities over time. In this chapter, we will explore the history of regenerative agriculture, starting with traditional agricultural practices, followed by the rise of industrial agriculture, and ending with the emergence of regenerative agriculture as a response to the negative impacts of industrial agriculture and the need for a more sustainable food system. By understanding the history of regenerative agriculture, we can gain insights into the challenges and opportunities ahead as we work towards a more sustainable and equitable food system.

Traditional Agricultural Practices

Traditional agricultural practices are those that have been passed down through generations and are rooted in the knowledge and wisdom of local communities. These practices are often closely tied to a given community's cultural and social traditions. As a result, they can vary widely depending on the region and the specific needs and conditions of the farm.

One common feature of traditional agricultural practices is using natural inputs, such as compost and animal manure, to enhance soil health and fertility. Compost is made from organic material that has been decomposed and is rich in nutrients and microbial life, which can improve the structure, biology, and nutrient content of the soil. Similarly, animal manure can be a valuable source of nutrients for the soil and provide organic matter that can improve the structure and water-holding capacity of the soil.

In addition to using natural inputs, traditional agricultural practices often involve techniques such as crop rotation and intercropping to enhance soil fertility and productivity. Crop rotation involves planting different crops in a specific order on the same piece of land. In contrast, intercropping involves the planting of different crops near each other. These practices can help improve soil health and fertility by adding nutrients and organic matter to the soil and reducing pests and diseases.

Overall, traditional agricultural practices offer a valuable source of knowledge and wisdom for modern agriculture, as they are based on the principles of sustainability and the understanding that the soil's health is essential for producing high-quality crops. By incorporating these practices into modern agriculture systems, it is possible to create a more sustainable and resilient food system that benefits the environment and the people who depend on it.

The Rise of Industrial Agriculture

We can trace the rise of industrial agriculture back to the post-World War II era when technological advances and the increasing demand for food led to the adoption of more intensive and mechanized forms of farming. This approach was based on synthetic inputs such as pesticides and fertilizers, as well as monoculture, the cultivation of a single crop over a large area. While industrial agriculture has contributed to increased food production, it has also negatively impacted the environment and the food system's sustainability. These impacts include soil degradation, water pollution, and biodiversity loss.

One of the main drivers of the rise of industrial agriculture was the increasing demand for food due to population growth and urbanization.

As more people moved from rural areas to cities, the demand for food increased, leading to the adoption of more efficient and productive farming practices. In addition, the development of new technologies, such as chemical fertilizers and pesticides, also played a role in the rise of industrial agriculture, as these products promised to increase crop yields and reduce the need for labor.

However, the adoption of industrial agriculture practices has also had negative consequences. The reliance on synthetic inputs such as pesticides and fertilizers can lead to soil degradation, as these products can strip the soil of nutrients and disrupt the natural balance of the ecosystem. In addition, the use of monoculture can lead to the loss of biodiversity, as it relies on the cultivation of a single crop over a large area rather than a diverse array of crops. This can make the farm system more vulnerable to pests and diseases, as well as the impacts of climate change.

Overall, the rise of industrial agriculture has contributed to increased food production. Still, it has also negatively impacted the environment and the food system's sustainability. To address these challenges, we must shift towards more sustainable and regenerative practices, such as those in regenerative agriculture.

The Emergence of Regenerative Agriculture

Regenerative agriculture is a holistic approach to farming and land management that aims to restore and enhance the health and fertility of the soil, as well as the biodiversity of the ecosystem. It is based on sustainability, resilience, and regenerative capacity principles. It seeks to create a closed-loop system in which waste is minimized and resources are used efficiently. Many different practices and techniques fall under the umbrella of regenerative agriculture, including cover cropping, composting, crop rotation, intercropping, agroforestry, and managed grazing, among others. These practices are often integrated and customized to suit the specific needs and conditions of a given farm or region.

One of the key goals of regenerative agriculture is to improve the health and fertility of the soil. Soil is a vital natural resource that provides the foundation for all plant growth and is essential for producing high-quality crops. Unfortunately, many modern farming practices have degraded the soil, leading to decreased crop yields and increased reliance on synthetic inputs such as fertilizers and pesticides. Regenerative agriculture seeks to reverse this trend by using practices that enhance the soil's structure, biology, and nutrient content, such as cover cropping, composting, and reducing tillage.

In addition to improving soil health, regenerative agriculture also seeks to conserve water and reduce erosion, enhance biodiversity, and integrate livestock in a way that benefits the ecosystem's overall health. These practices can help create a more sustainable and resilient food system that can adapt and thrive in the face of challenges such as climate change and population growth.

Overall, regenerative agriculture offers a promising approach to farming and land management that has the potential to create a more sustainable and equitable food system for the future. By incorporating these principles and practices into our farming and land management systems, we can work towards a brighter future for agriculture and the environment and create a world in which food is produced in a way that is environmentally responsible, socially just, and economically viable.

Chapter Summary

- The history of agriculture reflects changing needs and priorities over time, including the emergence of regenerative agriculture.
- Traditional agricultural practices are rooted in local knowledge and wisdom and often involve using natural inputs and techniques such as crop rotation and intercropping to enhance soil health and fertility.
- The rise of industrial agriculture was driven by population growth, urbanization, and technological advances but has negatively impacted the environment and sustainability.

- Industrial agriculture relies on synthetic inputs such as pesticides and fertilizers, monoculture, and large-scale mechanization.
- The negative impacts of industrial agriculture include soil degradation, water pollution, and biodiversity loss.
- The emergence of regenerative agriculture is a response to the negative impacts of industrial agriculture and the need for a more sustainable food system.
- Regenerative agriculture practices focus on soil health and fertility, water management, biodiversity, crop rotation and intercropping, and livestock management.
- Understanding the history of regenerative agriculture can provide insights into the challenges and opportunities of creating a more sustainable and equitable food system.

3

THE ENVIRONMENTAL BENEFITS OF
REGENERATIVE AGRICULTURE

Regenerative agriculture is not only a sustainable approach to farming but also offers numerous environmental benefits. In this chapter, we will explore the specific environmental benefits of regenerative agriculture, including soil carbon sequestration, water conservation, pesticide and fertilizer reduction, and biodiversity conservation. Furthermore, by understanding how these practices can support the health and resilience of the ecosystem, we can gain insights into the potential for regenerative agriculture to create a more sustainable and equitable food system for the future.

Soil Carbon Sequestration

Soil carbon sequestration is a critical aspect of regenerative agriculture, as it addresses the issue of excess carbon dioxide in the atmosphere, which is a major contributor to global warming. Carbon dioxide is a greenhouse gas that traps heat in the atmosphere, increasing the Earth's surface temperature. While carbon dioxide is a naturally occurring gas essential for life on Earth, burning fossil fuels and other human activities have contributed to an excess of this gas in the atmosphere, leading to climate change.

One way to address this excess of carbon dioxide is through carbon sequestration, which captures and stores carbon dioxide in the form of organic matter in the soil. Soil is a natural sponge that can absorb and store carbon dioxide, and regenerative agriculture practices can help to increase the amount of carbon stored in the soil. This is achieved through practices such as cover cropping and applying compost, which add organic matter to the soil and improve its structure and fertility.

In addition to the environmental benefits of carbon sequestration, there are also economic benefits for farmers. Carbon sequestration can increase the value of soil, as it is a valuable resource that we can trade on carbon markets. This can provide an additional source of income for farmers and encourage the adoption of regenerative agriculture practices.

Overall, soil carbon sequestration is a key aspect of regenerative agriculture, as it offers a way to address the issue of climate change and improve the health and fertility of the soil. By focusing on carbon sequestration and other regenerative agriculture practices, we can work towards a more sustainable and resilient food system that benefits the environment and the people who depend on it.

Water Conservation

Water conservation is essential for the long-term sustainability of agriculture, as it helps to reduce the demand for limited water resources and the associated environmental impacts of irrigation. In many parts of the world, water scarcity is a growing concern. In addition, population growth and climate change are increasing water demand. At the same time, supplies are being strained by drought and other factors. By adopting regenerative agriculture practices that prioritize water conservation, farmers can help to mitigate these challenges and create a more resilient and sustainable food system.

One of the key strategies for water conservation in regenerative agriculture is the use of mulching. Mulching involves the application of a layer of organic material, such as straw or wood chips, to the surface of the soil. This helps to retain moisture in the soil and reduce evaporation, as well as reducing erosion and the need for irrigation. Mulching can also improve the structure and fertility of the soil, as the organic material breaks down and adds nutrients and organic matter to the soil.

Contour planting is another water conservation strategy we can use in regenerative agriculture. This involves planting crops on the contour lines of a slope rather than in straight rows to reduce erosion and improve water use efficiency. By planting on the contour lines, water can flow down the slope in a more controlled manner, reducing erosion and the loss of soil and nutrients. Contour planting can also help improve the farm's overall productivity, as it maximizes the use of available water resources.

Using permeable surfaces is another water conservation strategy we can implement in regenerative agriculture. Permeable surfaces allow water to pass through them rather than running off or being absorbed into the ground. As a result, these surfaces can be used in driveways, paths, and patios. In addition, they can help reduce erosion and the need for irrigation by allowing water to infiltrate the soil. This can be particularly useful in scarce water, as it allows farmers to make the most efficient use of available water resources.

Rainwater harvesting is another key strategy for water conservation in regenerative agriculture. By collecting and storing rainwater, farmers can have a reliable source of water for their crops, reducing the need for irrigation and other external water sources. This can be particularly important in regions with scarce water or drought conditions. We can use various systems for rainwater harvesting, including simple systems such as cisterns or barrels that collect rainwater from rooftops to more complex systems that involve underground reservoirs or pond systems. In addition to providing a water source for irrigation, rainwater harvesting can also help reduce the demand for municipal water systems and improve the farm's overall efficiency of water use.

Greywater reuse is another water conservation strategy we can implement in regenerative agriculture. Greywater is wastewater from household activities such as washing dishes, laundry, and showers. It can be collected and treated for reuse in irrigation. This helps reduce the demand for freshwater, a limited and valuable resource. By using greywater in irrigation, farmers can save water and reduce the impact of their farming practices on the environment. However, it is important to properly treat and filter greywater to ensure that it is safe for irrigation and does not pose any health risks.

Overall, water conservation is an important aspect of regenerative agriculture, as it helps to conserve this vital resource and improve the efficiency of water use on the farm. Furthermore, by implementing strategies such as rainwater harvesting and greywater reuse, farmers can reduce their reliance on external water sources and contribute to the overall sustainability of the food system.

Pesticide and Fertilizer Reduction

One of the main goals of regenerative agriculture is to minimize synthetic inputs, such as pesticides and fertilizers, which can negatively impact the environment and human health. These inputs can pollute water sources and contribute to biodiversity loss, as they can harm beneficial insects and other wildlife. In contrast, regenerative agriculture practices aim to improve the soil's health and fertility through natural inputs and techniques such as crop rotation and intercropping. By enhancing the biology and nutrient content of the soil, regenerative agriculture can help to reduce the need for synthetic inputs and create a more sustainable and resilient food system.

Using natural inputs, such as compost and animal manure, can help improve soil health and fertility by adding nutrients and organic matter to the soil. These inputs also support the growth of beneficial microorganisms, which can improve the structure and biological diversity of the soil. In addition, crop rotation and intercropping can help to improve soil fertility and productivity by adding nutrients and organic matter, as well as reducing pests and diseases. By adopting these practices, farmers can create a more sustainable and resilient food system that can withstand the impacts of climate change and other challenges.

Overall, reducing synthetic inputs, such as pesticides and fertilizers, is a key goal of regenerative agriculture. Instead, by adopting practices that enhance soil health and fertility, farmers can create a more sustainable and resilient food system that benefits the environment and the people who depend on it.

Biodiversity Conservation

Regenerative agriculture practices support farm biodiversity and contribute to biodiversity conservation in the broader landscape. This is because regenerative agriculture systems often rely on a diverse range of crops and animals rather than monoculture, which can help to support a greater variety of species. In addition, regenerative agriculture practices such as cover cropping and agroforestry can provide habitat and food for wildlife, further contributing to biodiversity conservation.

The importance of biodiversity goes beyond the farm, as it plays a vital role in the overall health and functioning of the ecosystem. Biodiversity helps to maintain the balance of nature, as different species interact and support each other in various ways. For example, pollinators such as bees and butterflies play a critical role in the reproduction of many plants. At the same time, predators help to control pest populations. By promoting biodiversity on the farm, regenerative agriculture can help support the ecosystem's overall health and resilience.

Furthermore, biodiversity conservation is important to climate change adaptation and mitigation. As the climate changes, species will need to adapt to survive. Biodiversity can increase the resilience of ecosystems to climate change, as a diverse range of species is more likely to be able to adapt to changing conditions. By promoting biodiversity, regenerative agriculture can help increase the ecosystem's resilience to the impacts of climate change.

In conclusion, biodiversity conservation is a key environmental benefit of regenerative agriculture. By promoting a diverse range of crops and animals and enhancing habitat for pollinators and other beneficial insects, regenerative agriculture can help increase farm biodiversity and the broader landscape. This, in turn, can help support the overall health and resilience of the ecosystem and contribute to climate change adaptation and mitigation.

In conclusion, regenerative agriculture offers a promising approach to farming and land management that has the potential to create a more sustainable and equitable food system for the future. By incorporating the principles and practices of regenerative agriculture, including soil health and fertility, water management, biodiversity conservation, and the reduction of synthetic inputs, farmers and land managers can work towards a brighter future for agriculture and the environment. In addition to the environmental benefits of regenerative agriculture, there are also economic benefits for farmers, as these practices can improve the productivity and profitability of their operations. By shifting towards regenerative agriculture, we can create a food system that is sustainable, resilient, and equitable for all.

Chapter Summary

- Regenerative agriculture has numerous environmental benefits, including soil carbon sequestration, water conservation, pesticide and fertilizer reduction, and biodiversity conservation.
- Soil carbon sequestration captures and stores carbon dioxide in the form of organic matter in the soil, which can help mitigate climate change and improve soil health and fertility.
- Water conservation in regenerative agriculture involves strategies such as mulching and contour planting to reduce the demand for irrigation and improve water use efficiency.
- Pesticide and fertilizer reduction in regenerative agriculture is achieved through cover cropping and intercropping, which can reduce the need for synthetic inputs and improve soil health and fertility.
- Biodiversity conservation in regenerative agriculture is achieved through cover cropping and intercropping practices, which promote a diverse ecosystem and can improve soil health and fertility.
- The environmental benefits of regenerative agriculture can contribute to a more sustainable and equitable food system.

- Regenerative agriculture practices work together to create a closed-loop system that minimizes waste and maximizes resource efficiency.
- By understanding the environmental benefits of regenerative agriculture, we can gain insights into the potential for this approach to create a more sustainable and resilient food system.

4

THE ECONOMIC BENEFITS OF REGENERATIVE AGRICULTURE

I n addition to its environmental benefits, regenerative agriculture offers numerous economic benefits for farmers and communities. This chapter will explore how regenerative agriculture can increase efficiency and productivity, reduce input costs, enhance profitability, and support community development. By understanding the economic benefits of these practices, we can gain insights into the potential for regenerative agriculture to create more sustainable and equitable food systems.

Increased Efficiency and Productivity

Regenerative agriculture practices can improve the efficiency of water use on the farm. By using strategies such as mulching, contour planting, and rainwater harvesting, farmers can reduce the need for irrigation and other external water sources while conserving this vital resource. This can reduce production costs and increase the farm's overall profitability.

In addition to increasing efficiency and productivity, regenerative agriculture practices can also improve the quality of the crops produced. By focusing on soil health and fertility, farmers can produce healthier and more nutrient-dense crops, which can fetch a higher price in the market. This can provide an additional source of income for the farm and help increase the business's overall profitability.

Overall, the increased efficiency and productivity resulting from regenerative agriculture practices can provide various economic benefits for farmers. By shifting towards these practices, farmers can improve the profitability of their operations while also reducing costs and improving the overall sustainability of their farming practices.

Reduced Input Costs

Regenerative agriculture practices can also reduce input costs by reducing the need for irrigation. Water is a vital resource for agriculture, and the cost of irrigation can be a significant burden for farmers, particularly in areas where water is scarce or subject to drought conditions. By adopting regenerative agriculture practices that prioritize water conservation, such as rainwater harvesting and greywater use, farmers can reduce their reliance on irrigation and save on water costs.

In addition to reducing the costs of synthetic inputs and irrigation, regenerative agriculture practices can also help reduce labor costs. For example, many regenerative agriculture practices, such as cover cropping and mulch, can help suppress weeds, reducing the need for manual labor in the field. This can help to lower labor costs and increase the efficiency and productivity of the farm.

Overall, the potential to reduce input costs is an important economic benefit of regenerative agriculture, as it can help improve the farm's profitability and make it more financially sustainable in the long run. In addition, by adopting these practices, farmers can reduce their production costs and increase their operations' efficiency and productivity.

Increased Profitability

Regenerative agriculture can not only provide economic benefits for farmers, but it can also contribute to the overall sustainability and resilience of the food system. By improving the health and fertility of the soil and reducing the reliance on synthetic inputs, regenerative agriculture practices can create a more sustainable and resilient food system that is better able to withstand the challenges of a changing climate and shifting markets.

One way regenerative agriculture can increase profitability for farmers is through selling high-quality, sustainably-produced foods. Consumers are increasingly seeking out foods produced using sustainable practices and are willing to pay a premium for these products. By adopting regenerative agriculture practices, farmers can differentiate their products and capture a larger market share, leading to increased profits.

In addition to the sale of high-quality foods, regenerative agriculture can also provide economic benefits through the sale of carbon credits. Carbon sequestration, one of the key environmental benefits of regenerative agriculture, can also generate income for farmers. By sequestering carbon in the soil, farmers can participate in carbon markets and sell carbon credits to offset the greenhouse gas emissions of other industries. This can provide an additional source of income for farmers and further encourage the adoption of regenerative agriculture practices.

Overall, regenerative agriculture offers numerous economic benefits for farmers, including increased efficiency and productivity, reduced input costs, and increased profitability. Moreover, by adopting these practices, farmers can not only improve the sustainability and resilience of their farms but also contribute to the overall sustainability and resilience of the food system.

Community Development

In addition to creating economic opportunities for farmers, regenerative agriculture can also support community development by creating more sustainable and equitable food systems. For example, local food production and distribution can help reduce the distance between where food is grown and consumed, reducing the carbon emissions associated with transportation and storage. This can also help to support the local economy, as farmers can sell their products directly to consumers in their community.

Regenerative agriculture can also support community development by promoting social and environmental justice. By prioritizing soil and ecosystem health, regenerative agriculture can help protect the environment and the rights of marginalized communities. For example, by reducing synthetic inputs such as pesticides and fertilizers, regenerative agriculture can help protect the health of farm workers and the surrounding community.

Overall, regenerative agriculture can play a vital role in supporting community development by creating economic opportunities, promoting sustainable and equitable food systems, and protecting the environment and the rights of marginalized communities. By focusing on these principles, it is possible to create more resilient and sustainable communities for the future.

In conclusion, regenerative agriculture offers numerous economic benefits for farmers and communities. By increasing efficiency and productivity, reducing input costs, and enhancing profitability, regenerative agriculture practices can help to create more sustainable and equitable food systems. In addition to these benefits, regenerative agriculture can also support community development by creating new jobs and revitalizing local food systems. As more and more farmers adopt these practices, the economic benefits of regenerative agriculture will become increasingly apparent, making it a viable and valuable option for the future of agriculture.

Chapter Summary

- Regenerative agriculture offers numerous economic benefits for farmers and communities, including increased efficiency and productivity, reduced input costs, enhanced profitability, and support for community development.
- Increased efficiency and productivity in regenerative agriculture are achieved through mulching and contour planting, which can reduce the need for irrigation and improve crop quality.

- Reduced input costs in regenerative agriculture are achieved through rainwater harvesting and greywater use, which reduce the need for irrigation and cover cropping, which reduces the need for labor.
- We can achieve increased profitability in regenerative agriculture by selling high-quality, sustainably-produced foods and carbon credits.
- Support for community development in regenerative agriculture is achieved through community-supported agriculture, which connects farmers and consumers, and agroforestry, which provides economic and social benefits for local communities.
- The economic benefits of regenerative agriculture can contribute to a more sustainable and equitable food system.
- By understanding the economic benefits of regenerative agriculture, we can gain insights into the potential for this approach to create a more sustainable and resilient food system.
- The economic benefits of regenerative agriculture work together with the environmental benefits to create a more sustainable and productive farming system.

5

THE SOCIAL BENEFITS OF REGENERATIVE AGRICULTURE

R egenerative agriculture not only offers environmental and economic benefits but also has the potential to enhance social well-being and contribute to creating more equitable food systems. In this chapter, we will explore the specific social benefits of regenerative agriculture, including improved public health, enhanced rural livelihoods, increased food security, and strengthened communities. By understanding how these practices can support social well-being, we can gain insights into the potential for regenerative agriculture to create a more sustainable and equitable food system for the future.

Improved Public Health

Regenerative agriculture has the potential to improve public health in several ways significantly. Firstly, by reducing the reliance on synthetic inputs such as pesticides and fertilizers, regenerative agriculture practices can help to protect both the environment and human health. These inputs are often associated with negative impacts on ecosystems and human health, including the contamination of water sources and the development of chronic diseases. By replacing these inputs with natural alternatives and emphasizing soil health, regenerative agriculture can help to create a healthier and more sustainable food system.

In addition to reducing the negative impacts of synthetic inputs, regenerative agriculture can improve dietary health by producing nutrient-rich, locally-grown foods. These foods are often fresher and more nutrient-dense than those grown using conventional methods, which can positively impact consumers' health. For example, research has shown that locally-grown, nutrient-rich foods can help to reduce the risk of diet-related diseases such as obesity and heart disease.

Furthermore, regenerative agriculture practices can also support the development of community gardens and other forms of local food production, which can help to increase access to healthy, affordable foods for underserved communities. This can be particularly important in areas where access to fresh, nutritious foods is limited, as it can help to address food insecurity and improve overall health outcomes.

Overall, the improved public health resulting from regenerative agriculture is an important social benefit of these practices. By reducing the negative impacts of synthetic inputs, promoting the production of nutrient-rich foods, and supporting local food systems, regenerative agriculture can help to create a healthier and more sustainable food system for all.

Enhanced Rural Livelihoods

Regenerative agriculture has the potential to significantly enhance rural livelihoods by providing economic opportunities for farmers and creating more sustainable and equitable food systems. By focusing on the health and fertility of the soil and reducing the reliance on synthetic inputs, regenerative agriculture practices can improve the efficiency and productivity of farming operations, leading to increased profitability for farmers. In addition, adopting regenerative agriculture practices can also reduce input costs, such as irrigation and labor, which can further improve the financial sustainability of farming operations.

In addition to the economic benefits of regenerative agriculture for farmers, these practices can also support the development of local food systems and enhance rural livelihoods in other ways. For example, by supporting local food production and distribution, regenerative agriculture can help to create more resilient and sustainable communities. This can be particularly important in rural areas where access to fresh, nutritious foods is limited, as it can help to address food insecurity and improve overall health outcomes.

Furthermore, regenerative agriculture practices can also create new economic opportunities for farmers, such as producing and selling high-quality, sustainably-grown foods. As consumers become increasingly interested in the sustainability of their food choices, a growing demand for products produced using regenerative agriculture practices is growing. As a result, farmers can differentiate their products by meeting this demand and capturing a larger market share, increasing profits.

Overall, the enhanced rural livelihoods resulting from regenerative agriculture are an important social benefit of these practices. By providing economic opportunities for farmers and supporting the development of local food systems, regenerative agriculture can help to create more sustainable and equitable communities.

Increased Food Security

Increasing food security is an important social benefit of regenerative agriculture. By improving the health and fertility of the soil, regenerative agriculture practices can lead to increased crop yields and enhanced food production. This can be particularly important in areas where food insecurity is a major issue, as it can help to address shortages and ensure that communities have access to a reliable source of nutritious food.

One way regenerative agriculture can increase food security is through techniques such as crop rotation and intercropping. These practices can help improve the farm's overall efficiency and productivity by increasing the diversity of crops grown and enhancing the health of the soil. Crop rotation, in particular, can help to reduce the risk of soil degradation and improve the overall sustainability of the farm, as it allows farmers to alternate between different crops and rest the soil between growing seasons. This can help to maintain soil health and fertility over time, leading to increased crop yields and improved food security.

In addition to increasing crop yields and improving the efficiency and productivity of the farm, regenerative agriculture practices can also help to increase food security by promoting local food systems. By supporting the production and distribution of locally-grown foods, regenerative agriculture can help to reduce the reliance on imported foods and create more resilient and sustainable communities. This can be particularly important in areas where access to fresh, nutritious foods is limited, as it can help to address food insecurity and improve overall health outcomes.

Overall, the increased food security resulting from regenerative agriculture is an important social benefit of these practices. By improving the soil's health and fertility, enhancing the farm's efficiency and productivity, and promoting local food systems, regenerative agriculture can help ensure that communities have access to a reliable source of nutritious food.

Strengthened Communities

Regenerative agriculture has the potential to significantly strengthen communities by promoting local food production and distribution and by supporting the development of more resilient and sustainable communities. In addition, by focusing on the health and fertility of the soil and reducing the reliance on synthetic inputs, regenerative agriculture practices can improve the efficiency and productivity of farming operations, leading to increased profitability for farmers. This can create new economic opportunities and support the development of local food systems, which can contribute to the overall well-being of communities.

In addition to the economic benefits of regenerative agriculture for farmers, these practices can also support the development of more resilient and sustainable communities in other ways. For example, by promoting local food production and distribution, regenerative agriculture can help to reduce the reliance on imported foods and create more self-sufficient communities. This can be particularly important in areas where access to fresh, nutritious foods is limited, as it can help to address food insecurity and improve overall health outcomes.

Furthermore, regenerative agriculture practices can also contribute to the social well-being of communities by providing opportunities for education and collaboration. For example, by supporting the development of community gardens and other forms of local food production, regenerative agriculture can create opportunities for people to come together and learn from one another. This can help to build social connections and promote a sense of community, which can contribute to overall well-being.

Overall, the strengthened communities resulting from regenerative agriculture are an important social benefit of these practices. By promoting local food production and distribution, supporting the development of more resilient and sustainable communities, and enhancing social well-being, regenerative agriculture can help to create more equitable and sustainable food systems.

In summary, regenerative agriculture offers numerous social benefits that can positively impact the environment, farmers, and consumers. By focusing on the health and fertility of the soil and reducing the reliance on synthetic inputs, regenerative agriculture practices can improve the efficiency and productivity of farming operations, leading to increased profitability for farmers and enhanced rural livelihoods. In addition, regenerative agriculture can improve public health by producing nutrient-rich, locally-grown foods and reducing the negative impacts of synthetic inputs. These practices can also support the development of local food systems and increase food security, helping to ensure that communities have access to a reliable source of nutritious food. Finally, regenerative agriculture can strengthen communities by promoting local food production and distribution and supporting the development of more resilient and sustainable communities.

Overall, by incorporating the principles of regenerative agriculture into farming and land management practices, it is possible to create a more sustainable and equitable food system that benefits all stakeholders. By prioritizing the health and fertility of the soil and by reducing the reliance on synthetic inputs, regenerative agriculture can help to create a food system that is more resilient, productive, and sustainable in the long run.

Chapter Summary

- Regenerative agriculture can improve public health by reducing the reliance on synthetic inputs, producing nutrient-rich, locally-grown foods, and supporting community gardens and local food systems.
- These practices can enhance rural livelihoods by providing economic opportunities for farmers, creating more sustainable and equitable food systems, and supporting local food production and distribution.
- Regenerative agriculture can increase food security by supporting local food systems and increasing the availability of nutrient-rich foods.
- These practices can also strengthen communities by promoting social connections, supporting cultural traditions, and fostering a sense of place and identity.
- To realize the social benefits of regenerative agriculture, we must adopt a holistic approach that considers the needs and priorities of different stakeholders, including farmers, consumers, and the broader community.
- Policy and regulatory frameworks can support the adoption of regenerative agriculture practices as an education and outreach effort.
- Public-private partnerships and other forms of collaboration can also be useful in promoting the adoption and scaling of regenerative agriculture practices.
- To achieve regenerative agriculture's social, environmental, and economic benefits, we must adopt a systems-level approach that considers the interconnectedness of different factors and stakeholders in the food system.

6

CHALLENGES AND BARRIERS TO IMPLEMENTING REGENERATIVE AGRICULTURE

Despite its numerous benefits, adopting regenerative agriculture practices can be challenging, and several barriers may prevent its widespread implementation. This chapter will explore some of the main challenges and barriers to implementing regenerative agriculture, including lack of knowledge and resources, limited market demand, regulatory barriers, and cultural and social barriers. By understanding these challenges and barriers, we can gain insights into the potential strategies and solutions for promoting the adoption of regenerative agriculture practices.

Lack of Knowledge and Resources

One of the main challenges to the widespread adoption of regenerative agriculture practices is the need for more knowledge and resources among farmers. Many farmers may need to become more familiar with these practices. In addition, they may need access to the necessary information and resources to implement them on their farm. This can be a significant barrier to the adoption of regenerative agriculture, as farmers may be hesitant to adopt new practices without the necessary support and guidance.

In addition to the lack of knowledge and resources, there may also be a need for more training and extension programs to support farmers adopting regenerative agriculture practices. These programs can provide valuable information and resources to farmers, helping them understand the benefits and challenges of regenerative agriculture and providing guidance on implementing these practices on their farms. Without access to these programs, farmers may be less likely to adopt regenerative agriculture practices, even if they are interested in doing so.

Furthermore, the lack of knowledge and resources can also challenge farmers interested in transitioning to regenerative agriculture but need help figuring out where to start. However, with access to the necessary information and resources, farmers may be able to develop a clear plan for implementing these practices on their farms, which can be a significant barrier to the adoption of regenerative agriculture.

The lack of knowledge and resources is a significant challenge to the widespread adoption of regenerative agriculture practices. By providing more information and resources to farmers and supporting the development of training and extension programs, it is possible to overcome this challenge and support the widespread adoption of regenerative agriculture practices.

Limited Market Demand

Limited market demand for sustainably-produced foods is another significant challenge to the widespread adoption of regenerative agriculture practices. While there is growing interest in sustainably-produced foods, more is needed to support the widespread adoption of regenerative agriculture practices. This can be a significant barrier to adopting regenerative agriculture. Farmers may be hesitant to invest in these practices if they are unsure of their ability to sell their products.

One way we can address this challenge is through developing marketing and branding strategies that highlight the benefits of sustainably-produced foods and help differentiate these products from conventionally-grown foods. In addition, promoting the sustainability and quality of sustainably-produced foods may increase consumer demand for these products and create more opportunities for farmers interested in adopting regenerative agriculture practices.

In addition to marketing and branding efforts, there may also be a need for policy changes and other interventions to support the development of a more robust market for sustainably-produced foods. For example, governments and other organizations may need to work to create incentives for the production and purchase of sustainably-grown foods or to support the development of distribution channels for these products. By addressing these challenges and creating a more supportive market environment for sustainably-produced foods, it may be possible to increase the demand for these products and support the widespread adoption of regenerative agriculture practices.

Regulatory Barriers

Regulatory barriers can significantly challenge the implementation of regenerative agriculture practices. For example, certain practices may be restricted by law, or financial incentives or subsidies may support the adoption of industrial agriculture practices rather than regenerative agriculture practices. These regulatory barriers can make it difficult for farmers to adopt regenerative agriculture practices, as they may be hesitant to invest in these practices if they are unsure of their legal status or the availability of financial support.

We can address regulatory barriers by developing policies and regulations supporting regenerative agriculture practices. For example, governments and other organizations may need to work to remove restrictions on these practices or to provide financial incentives or subsidies to farmers who are interested in adopting these practices. By creating a more supportive policy environment for regenerative agriculture, it may be possible to overcome regulatory barriers and encourage the widespread adoption of these practices.

In addition to policy interventions, there is also a need for education and outreach efforts to increase awareness of the benefits of regenerative agriculture and the challenges posed by regulatory barriers. By providing more information to farmers and other stakeholders about the benefits and challenges of these practices, it may be possible to build support for policy changes that support the adoption of regenerative agriculture practices.

Overall, regulatory barriers can significantly challenge the implementation of regenerative agriculture practices. By addressing these barriers through policy changes, education, and outreach efforts, it may be possible to create a more supportive environment for regenerative agriculture and encourage the widespread adoption of these practices. By removing restrictions on regenerative agriculture practices and providing financial incentives and subsidies to farmers interested in adopting these practices, governments and other organizations can help overcome regulatory barriers and support the development of more sustainable and equitable food systems.

Cultural and Social Barriers

Cultural and social barriers can also significantly challenge adopting regenerative agriculture practices. For example, traditional farming practices are deeply ingrained in certain cultures, and there may be resistance to change. This can be a significant barrier to adopting regenerative agriculture practices, as farmers may be hesitant to adopt new practices if they are unfamiliar or perceived as conflicting with traditional ways of farming.

In addition to cultural barriers, there may also be social and economic pressures that discourage the adoption of regenerative agriculture practices. For example, there may be a need to maximize short-term profits, which can lead farmers to prioritize efficiency and productivity over their farming operations' long-term health and sustainability. There may also be a need for more access to capital and other resources, making it difficult for farmers to invest in the equipment and infrastructure needed to implement regenerative agriculture practices.

To overcome these cultural and social barriers, engaging in education and outreach efforts that highlight the benefits of regenerative agriculture and address any concerns or misconceptions about these practices may be necessary. In addition, there may be a need for policy interventions and other support mechanisms to address the social and economic pressures that discourage the adoption of regenerative agriculture practices. However, by addressing these cultural and social barriers and creating a more supportive environment for regenerative agriculture, it may be possible to overcome these challenges and encourage the widespread adoption of these practices.

In summary, there are several challenges and barriers to implementing regenerative agriculture, including a lack of knowledge and resources, limited market demand, regulatory barriers, and cultural and social barriers. Understanding these challenges and barriers makes it possible to identify strategies and solutions for promoting the adoption of regenerative agriculture practices and creating more sustainable and equitable food systems.

Chapter Summary

- Lack of knowledge and resources can significantly challenge the widespread adoption of regenerative agriculture practices.
- Limited market demand for sustainably-produced foods can also be a barrier to adoption.
- Regulatory barriers, such as subsidies for synthetic inputs, can discourage the adoption of regenerative agriculture practices.
- Cultural and social barriers, such as traditional farming practices and skepticism towards new approaches, can also hinder adoption.
- Financial barriers, such as the initial investment required to implement regenerative agriculture practices, can challenge farmers.
- Access to land and other resources, such as water, can also be a barrier to adoption.

- Political and economic factors, such as trade policies and market forces, can impact the feasibility of adopting regenerative agriculture practices.
- There needs to be more research and data on the effectiveness of regenerative agriculture practices to make it easier for farmers to make informed decisions about adoption.

7

SOLUTIONS AND STRATEGIES FOR OVERCOMING CHALLENGES

D espite the numerous challenges and barriers to implementing regenerative agriculture, we can employ a range of solutions and strategies to overcome these challenges and promote the adoption of these practices. This chapter will explore some of the main solutions and strategies for overcoming challenges to regenerative agriculture, including education and training programs, research and development, policy and regulatory reform, and market development and support. By understanding these strategies and solutions, we can gain insights into the potential for promoting the adoption of regenerative agriculture practices and creating more sustainable and equitable food systems.

Education and Training Programs

Education and training programs are an important solution to the challenge of lack of knowledge and resources when adopting regenerative agriculture practices. Providing farmers with the necessary information and resources, as well as training in these practices, supports the adoption of regenerative agriculture and helps farmers to understand the benefits and challenges of these practices.

There are several types of education and training programs that can be effective in supporting the adoption of regenerative agriculture practices. For example, extension programs can provide valuable information and resources to farmers, as well as technical assistance and guidance on implementing these practices on their farms. In addition, online courses, workshops, and other forms of training can provide farmers with the knowledge and skills they need to understand and implement regenerative agriculture practices.

One of the key benefits of education and training programs is that they can provide farmers with the support they need to overcome the challenges and barriers to adopting regenerative agriculture practices. Providing access to information and resources, as well as guidance and technical assistance, can help farmers feel more confident and capable of implementing these practices on their farms. In addition, education and training programs can also help build a sense of community and support among farmers interested in adopting regenerative agriculture practices, which can be an important factor in helping these practices be more widely adopted.

Overall, education and training programs are a key solution to the challenge of lack of knowledge and resources when adopting regenerative agriculture practices. Providing farmers with the necessary information and resources, as well as training and technical assistance, can support adopting these practices and creating more sustainable and equitable food systems.

Research and Development

Research and development is a critical strategy for overcoming challenges to the widespread adoption of regenerative agriculture practices. Investing in research to improve our understanding of these practices and their impacts makes it possible to develop more effective and sustainable approaches to regenerative agriculture. This research can also help identify the most effective strategies and solutions for promoting the adoption of regenerative agriculture practices and can also provide valuable insights into the challenges and barriers that farmers and other stakeholders may face when implementing these practices.

Several types of research and development can be useful in addressing the challenges to regenerative agriculture. For example, research on these practices' environmental impacts can help identify the most sustainable approaches to regenerative agriculture and provide valuable information about the benefits of these practices for soil health, water conservation, and biodiversity. Research on the economic impacts of these practices can also be useful, as it can help identify the most cost-effective and profitable approaches to regenerative agriculture and provide insights into how these practices can support the development of more sustainable and equitable food systems.

In addition to research on regenerative agriculture's environmental and economic impacts, research on social and cultural issues can also be valuable. For example, research on the cultural and social barriers to adopting these practices can help identify strategies for overcoming these barriers and promoting the widespread adoption of these practices. Investing in research and development makes it possible to gain a more comprehensive understanding of the challenges and opportunities presented by regenerative agriculture and identify the most effective strategies for addressing these challenges and promoting the adoption of these practices.

Overall, research and development is a critical strategy for overcoming the challenges of the widespread adoption of regenerative agriculture practices. Investing in research to improve our understanding of these practices and their impacts makes it possible to develop more effective and sustainable approaches to regenerative agriculture and identify the most effective strategies and solutions for promoting the adoption of these practices.

Policy and Regulatory Reform

Policy and regulatory reform is a critical strategy for overcoming challenges to the widespread adoption of regenerative agriculture practices. By reforming policies and regulations that support industrial agriculture practices and discourage adopting regenerative agriculture practices, it is possible to create a more supportive environment for these practices. This may include changes to subsidies, incentives, and other financial support mechanisms, as well as changes to laws and regulations that align with the principles of regenerative agriculture.

One of the key benefits of the policy and regulatory reform is that it can create a more level playing field for regenerative agriculture practices, making it easier for farmers to adopt and compete in the marketplace. For example, by providing financial incentives or subsidies to farmers who adopt regenerative agriculture practices, it is possible to make these practices more attractive and economically viable for farmers. In addition, by reforming regulations that support industrial agriculture practices and discourage the adoption of regenerative agriculture practices, it is possible to create a more supportive environment for these practices, encouraging more farmers to adopt them.

Policy and regulatory reform can also include the development of standards and certification programs that recognize and reward the adoption of regenerative agriculture practices. By creating standards and certification programs that reflect the principles of regenerative agriculture, it is possible to create a more transparent and accountable system for recognizing and rewarding the adoption of these practices. This can create a more supportive environment for regenerative agriculture and encourage more farmers to adopt these practices.

Overall, policy and regulatory reform is a key strategy for overcoming challenges to the widespread adoption of regenerative agriculture practices. By reforming policies and regulations that support industrial agriculture practices and discourage the adoption of regenerative agriculture practices, it is possible to create a more supportive environment for these practices and create the conditions for them to thrive and become more widely adopted.

Market Development and Support

Market development and support are crucial strategies for overcoming challenges to the widespread adoption of regenerative agriculture practices. Increasing the demand for sustainably-produced foods makes it possible to create a more supportive environment for these practices and encourage more farmers to adopt them.

Several strategies can effectively increase the market demand for sustainably-produced foods and support the adoption of regenerative agriculture practices. For example, marketing and labeling initiatives can educate consumers about these foods' benefits and the practices used to produce them and create a more attractive and appealing market for these products. Direct-to-consumer sales, such as farmers' markets and community-supported agriculture (CSA) programs, can also increase the demand for sustainably-produced foods and support the adoption of regenerative agriculture practices. In addition, developing local and regional food systems can create more opportunities for farmers to sell their sustainably-produced foods and can help to increase the market demand for these products.

Increasing the market demand for sustainably-produced foods makes it possible to create economic incentives for farmers to adopt regenerative agriculture practices. By providing farmers with more opportunities to sell their sustainably-produced foods and creating a more attractive and appealing market for these products, it is possible to encourage more farmers to adopt these practices and create more sustainable and equitable food systems.

Overall, market development and support are crucial strategies for overcoming challenges to the widespread adoption of regenerative agriculture practices. Increasing the market demand for sustainably-produced foods makes it possible to create economic incentives for farmers to adopt these practices and create more sustainable and equitable food systems.

In conclusion, we can employ various solutions and strategies to overcome the challenges and barriers to implementing regenerative agriculture practices. These include education and training programs, research and development, policy and regulatory reform, and market development and support. Understanding these strategies and solutions makes it possible to identify the most effective approaches for promoting the adoption of regenerative agriculture practices and creating more sustainable and equitable food systems.

Chapter Summary

- Education and training programs are a key solution to the lack of knowledge and resources when adopting regenerative agriculture practices.
- Research and development is a critical strategy for improving our understanding of regenerative agriculture practices and developing more effective approaches.
- Policy and regulatory reform can help to create a more supportive environment for regenerative agriculture by promoting sustainable farming practices and supporting farmers interested in adopting these practices.
- Market development and support can help create a more robust market for sustainably-produced foods, increasing demand for these products and supporting the adoption of regenerative agriculture practices.
- Collaboration and partnerships can help build a sense of community and support among farmers and other stakeholders interested in regenerative agriculture, which can be important in promoting and adopting these practices.
- Financing and investment can help provide the necessary resources for farmers to adopt regenerative agriculture practices and support the development of more sustainable and equitable food systems.
- Public awareness and outreach can help increase understanding of regenerative agriculture's benefits and promote the adoption of these practices by a wider audience.
- Government and other policymakers have an important role in promoting the adoption of regenerative agriculture practices and creating a more sustainable and equitable food system.

8

THE FUTURE OF REGENERATIVE
AGRICULTURE

As the global population continues to grow and the impacts of climate change become more apparent, it is essential to consider regenerative agriculture's role in addressing these challenges and creating a more sustainable and equitable food system. In this chapter, we will explore the potential of regenerative agriculture to address climate change, feed the world, and support government and policy efforts to promote these practices.

The Role of Regenerative Agriculture in Addressing Climate Change

Regenerative agriculture has a significant role in addressing the challenges of climate change. One of the key ways in which these practices can contribute to climate change mitigation is through the sequestration of carbon in the soil. Practices such as using cover crops and applying compost can help increase the amount of organic matter in the soil, leading to increased carbon sequestration. This not only helps to mitigate the impacts of climate change but also improves the health and fertility of the soil.

In addition to carbon sequestration, regenerative agriculture practices can contribute to climate change mitigation by reducing greenhouse gas emissions. By reducing the need for synthetic inputs such as pesticides and fertilizers, regenerative agriculture practices can help reduce agriculture's environmental impacts and decrease the release of greenhouse gases. For example, using cover crops and compost can help increase soil health and fertility, reducing the need for synthetic inputs and the emissions associated with their production and use.

Regenerative agriculture practices can also contribute to climate change adaptation by improving the resilience of agricultural systems. By enhancing the health and fertility of the soil, these practices can help to improve the productivity and sustainability of agricultural systems and increase their ability to withstand the impacts of a changing climate. For example, regenerative agriculture practices that focus on soil health and water conservation can help to improve the resilience of agricultural systems in the face of drought and other extreme weather events.

Overall, regenerative agriculture has a significant role in addressing the challenges of climate change. By improving the health and fertility of the soil, reducing greenhouse gas emissions, and enhancing the resilience of agricultural systems, these practices can contribute to creating a more sustainable and equitable food system that is better able to withstand the impacts of a changing climate.

The Potential for Regenerative Agriculture to Feed the World

The potential for regenerative agriculture to feed the world is an important aspect of the future of these practices. While industrial agriculture has succeeded in increasing food production, it has also contributed to environmental degradation and biodiversity loss. In contrast, regenerative agriculture practices can increase food production sustainably and equitably.

One of the key ways in which regenerative agriculture can increase food production is by improving soil health and fertility. By adopting practices such as using cover crops and compost, farmers can improve the quality and fertility of the soil, leading to increased crop yields and improved food security. In addition, regenerative agriculture practices that focus on soil health and fertility can also enhance the resilience of agricultural systems, making them more resistant to the impacts of extreme weather events and other challenges.

Another way in which regenerative agriculture can increase food production is through the conservation of water. By adopting practices such as rainwater harvesting and greywater, farmers can reduce their reliance on irrigation and other external water sources, leading to increased efficiency and productivity. This can help to improve food security and increase the sustainability of agricultural systems.

Overall, the potential for regenerative agriculture to feed the world is significant. Improving soil health and fertility, conserving water, and enhancing ecosystem resilience can increase food production sustainably and equitably. By adopting regenerative agriculture practices, farmers can contribute to creating a more sustainable and equitable food system that can better feed the world's growing population.

The Role of Government and Policy in Supporting Regenerative Agriculture

The role of government and policy in supporting regenerative agriculture is critical to the widespread adoption and scaling of these practices. Governments and policymakers have several tools to support the adoption and scaling of regenerative agriculture practices, including education and training programs, research and development initiatives, and policy and regulatory reforms.

Education and training programs can be important tools for supporting the adoption of regenerative agriculture practices. Providing farmers with the necessary information and resources, as well as training in these practices, supports the adoption of these practices and creates a more supportive environment for regenerative agriculture. In addition, extension programs and other forms of technical assistance can also effectively provide farmers with the support they need to implement regenerative agriculture practices on their farms.

Research and development initiatives are another important tools for supporting regenerative agriculture adoption. Investing in research to improve our understanding of these practices and their impacts makes it possible to develop more effective and sustainable approaches to regenerative agriculture. This research can also help to identify the most effective strategies and solutions for promoting the adoption of regenerative agriculture practices.

Policy and regulatory reform is another key strategy for supporting the adoption of regenerative agriculture practices. By reforming policies and regulations that support industrial agriculture practices and discourage adopting regenerative agriculture practices, it is possible to create a more supportive environment for these practices. This may include changes to subsidies, incentives, and other financial support mechanisms, as well as changes to laws and regulations that align with the principles of regenerative agriculture. Policy and regulatory reform can also include the development of standards and certification programs that recognize and reward the adoption of regenerative agriculture practices.

Finally, governments and policymakers can support the development of alternative food systems, such as local and regional food systems, based on the principles of regenerative agriculture. By creating more opportunities for farmers to sell their sustainably-produced foods and by supporting the development of more resilient and sustainable communities, it is possible to create a more supportive environment for regenerative agriculture practices and contribute to creating a more sustainable and equitable food system. Local and regional food systems can increase the market demand for sustainably-produced foods, create economic opportunities for farmers, and enhance the resilience of communities. By investing in the development of alternative food systems, governments and policymakers can help create the necessary conditions for the adoption and scaling of regenerative agriculture practices and contribute to creating a more sustainable and equitable food system.

The future of regenerative agriculture is bright, with the potential to address some of the most pressing challenges facing the world today, including climate change and food security. Focusing on soil health and fertility, water conservation, biodiversity, and other principles of regenerative agriculture makes it possible to create a more sustainable and equitable food system that benefits the environment, farmers, and consumers.

The role of government and policy in supporting regenerative agriculture is also crucial, as it can help to create a more supportive environment for these practices and accelerate their adoption. While there are challenges and barriers to the implementation of regenerative agriculture, there is also a range of solutions and strategies that we can employ to overcome these challenges. By working together and sharing knowledge and resources, it is possible to create a brighter future for regenerative agriculture and the food system.

Chapter Summary

- Regenerative agriculture can address climate change by sequestering carbon in the soil, reducing greenhouse gas emissions, and enhancing the resilience of agricultural systems.
- Regenerative agriculture has the potential to increase food production sustainably and equitably by improving soil health and fertility, conserving water, and promoting biodiversity.
- Governments and policymakers can support the adoption of regenerative agriculture practices through financial incentives, research and development, and regulatory reform.
- International organizations and NGOs can support the adoption of regenerative agriculture by promoting education and training programs, conducting research and development, and advocating for policy reform.
- Consumers can support the adoption of regenerative agriculture by purchasing sustainably-produced foods, participating in community-supported agriculture programs, and advocating for policy reform.
- The private sector can support the adoption of regenerative agriculture by investing in research and development, promoting sustainable supply chains, and advocating for policy reform.

- Land conservation organizations can support the adoption of regenerative agriculture by promoting education and training programs, conducting research and development, and advocating for policy reform.
- Universities and other educational institutions can support the adoption of regenerative agriculture by conducting research, promoting education and training programs, and advocating for policy reform.

CONCLUSION

I n this book, we have explored the principles and benefits of regenerative agriculture and the challenges and barriers to its implementation. We have examined the role of soil health and fertility, water management, biodiversity, and livestock management in regenerative agriculture and how these practices can improve crop yields, reduce the need for synthetic inputs, enhance ecosystem resilience, and create more sustainable communities. We have also highlighted regenerative agriculture's economic and social benefits, including increased efficiency and productivity, reduced input costs, increased profitability, improved public health, enhanced rural livelihoods, increased food security, and strengthened communities.

We have also discussed the challenges and barriers to implementing regenerative agriculture, including lack of knowledge and resources, limited market demand, regulatory barriers, and cultural and social barriers. Finally, we have explored solutions and strategies for overcoming these challenges, including education and training programs, research and development, policy and regulatory reform, and market development and support.

In the final chapter, we considered the future of regenerative agriculture and the role it can play in addressing climate change and feeding the world. We also explored the importance of government and policy in supporting regenerative agriculture and the potential for these practices to create a more sustainable and equitable food system.

Overall, it is clear that regenerative agriculture offers a promising and holistic approach to farming and land management that has the potential to address many of the challenges facing the food system today. By adopting these practices and working together to overcome challenges and barriers, it is possible to create a brighter future for regenerative agriculture and the food system.

INTRODUCTION TO SOIL SCIENCE

FROM FORMATION AND CLASSIFICATION TO PHYSICAL, CHEMICAL, AND BIOLOGICAL PROPERTIES, FERTILITY AND NUTRIENT MANAGEMENT, AND EROSION AND CONSERVATION

INTRODUCTION

Soil science studies soil as a natural resource on the Earth's surface. It is a multi-disciplinary field that encompasses the physical, chemical, biological, and ecological properties and processes in soil and is essential for understanding and managing this vital resource. **Soil is a complex and dynamic medium that supports the growth of plants.** It is crucial for maintaining the health of ecosystems. It is also an important resource for agriculture, forestry, and other land-use activities. This chapter will introduce soil science, including its definition, importance, history, and development field.

Definition of Soil Science

Soil science is a diverse and interdisciplinary field encompassing various topics and applications. It is a required field of study, as the soil is a vital resource for the growth of plants and other land use activities and is a critical component of ecosystems worldwide. Soil scientists work to understand the properties and processes of soil and how they interact to develop strategies for managing and sustainably conserving soil resources.

Soil science involves the study of the physical, chemical, and biological properties of soil, as well as the ecological processes that occur within the soil. Physical properties of soil include things like texture, structure, and density, which can all impact soil's ability to support plants' growth. The chemical properties of soil include pH, nutrients, and organic matter, which can also influence plant growth and soil fertility. The biological properties of soil refer to the microorganisms and other organisms that live within the soil and their role in soil health and fertility. Finally, ecological processes within soil involve the interactions between soil's physical, chemical, and biological properties and how they impact the overall functioning of soil ecosystems.

Soil science also involves studying how soil forms and changes over time. Soil is created through soil formation, which involves interacting with various physical, chemical, and biological factors. These factors include climate, topography, parent material, and vegetation, which can all influence the characteristics and properties of soil. Soil scientists work to understand these factors and how they interact with one another to understand better the process of soil formation and the factors that influence soil properties.

Soil science has many applications, including agriculture, forestry and landscaping, environmental restoration, and urban and industrial settings. For example, in agriculture, soil scientists work to understand the factors that influence soil productivity and fertility and develop strategies for managing soil to maximize crop yields and promote soil health. In forestry and landscaping, soil scientists study the characteristics and needs of different soil types and develop techniques for managing soil in these settings to promote the growth of trees and other plants. In environmental restoration, soil scientists work to rehabilitate degraded soils and restore ecosystem function. And in urban and industrial settings, soil scientists work to understand the unique challenges and characteristics of soil in these environments and develop techniques for managing soil to promote the growth of plants and other land-use activities while conserving soil resources.

Overall, soil science is a vital field of study for understanding and sustainably managing soil resources. By studying the physical, chemical, and biological properties of soil, and the ecological processes that occur within the soil, soil scientists can develop strategies for managing and conserving soil resources and make important contributions to a wide range of fields.

Importance of Soil Science

Soil is a vital resource that is essential for supporting life on Earth. It is a complex and dynamic system comprising various physical, chemical, and biological components. **It plays a critical role in supporting the growth of plants and other land-use activities.** Soil is also a vital component of *ecosystems*. It provides a habitat for a diverse array of microorganisms and other organisms. In addition, it plays a role in the cycling of nutrients and water.

The importance of soil science cannot be overstated, as it is essential for understanding and sustainably managing soil resources. Soil science is a field of study that examines soil's physical, chemical, biological, and ecological properties and processes and how they interact with one another. In addition, soil scientists work to understand how soil forms and changes over time and can be managed and conserved to promote its long-term health and productivity.

Soil science is important for various applications, including agriculture, forestry, and other land use activities. In agriculture, soil science is essential for understanding the factors that influence soil productivity and fertility and for developing strategies for managing soil to maximize crop yields and promote soil health. In forestry, soil science is important for understanding the characteristics and needs of different types of soil and for developing techniques for managing soil to promote the growth of trees and other plants. Soil science is also important for other land use activities, such as landscaping, environmental restoration, and urban and industrial settings. It provides the knowledge and tools to understand and manage soil resources in these contexts.

Soil science is also important for understanding soil's role in maintaining ecosystems' health. Soil is a vital component of ecosystems and plays a role in cycling nutrients, water, and other resources. By studying the ecological processes that occur within the soil, soil scientists can understand how soil impacts the overall functioning of ecosystems and how it can be managed to promote ecosystem health and stability.

In summary, soil science is a critical field of study essential for understanding and sustainably managing soil resources. By studying the physical, chemical, and biological properties of soil, and the ecological processes that occur within the soil, soil scientists can develop strategies for managing and conserving soil resources and make important contributions to a wide range of fields.

History and Development of Soil Science

We can trace the history of soil science back to ancient civilizations, which recognized soil's importance for agriculture and other land use activities. The ancient Egyptians, for example, used irrigation and other techniques to improve soil fertility and support the growth of crops. Similarly, the ancient Greeks and Romans also recognized the importance of soil for agriculture and developed techniques for improving soil fertility and productivity.

It was in the *19th,* and *20th centuries*, however, that soil science emerged as a distinct field of study. During this time, advances in chemistry, biology, and other fields helped us better understand soil's complex properties and processes. These advances included the development of new analytical techniques for studying soil, as well as the discovery of new elements and compounds that are important for soil health and fertility.

The development of soil science also coincided with a growing recognition of the importance of soil conservation. As the human population grew and land use activities increased, there was a growing concern about the impact of these activities on soil resources. This led to the developing of new techniques and strategies for managing and conserving soil resources, such as soil conservation practices and sustainable land use practices.

Today, soil science continues to evolve and expand, with new research and innovations helping to further our understanding of this vital resource. For example, soil scientists can now use a wide range of **analyt-**

ical techniques to study soil and have made important advances in soil fertility and nutrient management, soil erosion and conservation, and soil management in different land use contexts. In addition, soil science continues to be an important field of study in the face of global challenges such as climate change and environmental degradation, as soil plays a key role in maintaining the health and stability of ecosystems.

In summary, soil science is an important field that plays a crucial role in understanding and managing soil resources. By studying the physical, chemical, biological, and ecological properties and processes of soil, soil scientists can develop techniques for conserving and improving soil quality and advise on the sustainable use of this vital resource. Furthermore, the field of soil science is constantly evolving, with new research and innovations emerging all the time. As such, it is important to continue investing in soil science research and education to ensure our soil resources' long-term health and productivity.

1

SOIL FORMATION AND CLASSIFICATION

S oil formation is the process by which soil is created and develops over time. It is a complex process influenced by various factors, including climate, vegetation, geology, and the actions of living organisms. Soil classification is the process of organizing and categorizing soils based on their physical and chemical properties. This is important for understanding different soils' characteristics and potential uses and developing appropriate management practices. In this chapter, we will explore the processes of soil formation, the factors that influence it, and the classification of soils based on their physical and chemical properties.

Processes of Soil Formation

Soil formation, also known as pedogenesis, is the process by which soil is created and develops over time. It is a complex process that involves the interaction of various physical, chemical, and biological factors. These factors interact with one another to create soil, and the specific combination of these factors influences soil characteristics in a particular location.

Physical factors influencing soil formation include climate, topography, and the actions of water, wind, and other weathering agents. Climate, for example, can influence the type of soil that forms in a particular location, as different types of soil are better suited to different climatic conditions. Topography, or the shape and slope of the land, can also influence soil formation, as different types of soil are better suited to different topographic conditions. Water, wind, and other weathering agents also play a role in soil formation, as they help to break down and erode the materials that makeup soil.

Chemical factors that influence soil formation include the presence of minerals and organic matter and the effects of acidity and pH. Different minerals can contribute to soil characteristics, such as its texture, structure, and nutrient content. Organic matter, made up of decomposed plant and animal material, is also an important component of soil, as it helps improve soil structure and fertility. Acidic or basic conditions, which are determined by the pH of the soil, can also influence soil formation, as certain types of soil are better suited to different pH conditions.

Biological factors influencing soil formation include the actions of living organisms, such as plants, animals, and microorganisms. Plants play a critical role in soil formation, as they help to break down and decompose organic matter and contribute to the nutrient content of the soil. Animals and microorganisms also play a role in soil formation, as they help to break down and decompose organic matter and contribute to the nutrient-cycling process within the soil.

Overall, soil formation is a complex process involving various physical, chemical, and biological factors. The specific combination of these factors in a particular location determines the characteristics and properties of the soil that forms in that location.

Factors that Influence Soil Formation

Many factors can influence soil formation, including climate, vegetation, geology, and the actions of living organisms. Understanding these factors is important for understanding how soil forms and changes over time and developing strategies for managing and conserving soil resources.

Climate is one of the most important factors influencing soil formation, as it determines the amount of rainfall, temperature, and other weather patterns in an area. Different soil types are better suited to different climatic conditions, and the specific combination of climatic conditions in a particular location can influence soil characteristics. For example, soils in areas with high levels of rainfall tend to be more acidic, as rainwater can leach minerals from the soil and lower its pH. Similarly, soils in areas with high levels of sunlight and heat tend to be more alkaline, as the heat and sunlight can break down organic matter and increase the pH of the soil.

Vegetation, including plants and trees, can also influence soil formation, as they contribute organic matter and nutrients to the soil and help to protect it from erosion. Different plants and trees have different effects on soil, depending on their roots, the amount of leaf litter they produce, and the types of nutrients they contribute to the soil. For example, trees with deep root systems can help to anchor the soil and prevent erosion. In contrast, plants with shallow root systems may contribute more organic matter and nutrients to the soil.

Geology, including the type and composition of rock and minerals in an area, can also influence soil formation, as these materials can be broken down and weathered to form soil. In addition, the type and composition of rock and minerals in an area can influence the soil's characteristics, such as its texture, structure, and nutrient content.

The actions of living organisms, such as plants, animals, and microorganisms, can also influence soil formation, as they help to break down and decompose organic matter. **Microorganisms, such as bacteria and fungi, are important decomposers of organic matter and play a critical role in the nutrient-cycling process within the soil.** Animals, such as worms and insects, also contribute to the decomposition of organic matter and can help to improve soil structure and fertility.

Overall, many factors can influence soil formation. Therefore, understanding these factors is important for understanding how soil forms and changes over time and developing strategies for managing and conserving soil resources.

Classification of Soils Based on Physical and Chemical Properties

Soil classification is the process of organizing and categorizing soils based on their physical and chemical properties. This is important for understanding different soils' characteristics and potential uses and developing appropriate management practices. Some common factors used in soil classification include texture, structure, pH, nutrient content, and organic matter content.

Soil texture refers to the size and proportions of the various particles that make up soil, including sand, silt, and clay. Soil structure refers to how these particles are arranged and held together and can be influenced by factors such as the presence of organic matter and the actions of living organisms. pH, or the acidity and alkalinity of the soil, is another important factor in soil classification, as different plants are better suited to different pH conditions. Nutrient content, including the levels of essential nutrients such as nitrogen, phosphorus, and potassium, is also an important factor in soil classification, as it can influence the health and productivity of plants. Finally, organic matter content, which refers to the amount of decomposed plant and animal material in soil, can also influence soil classification, as it helps to improve soil structure and fertility.

There are several different soil classification systems have been developed over the years, including the United States Department of Agriculture (USDA) soil taxonomy, the World Reference Base for Soil Resources (WRB), and the International Union of Soil Sciences (IUSS) soil classification system. These systems use different criteria and approaches to categorize soils. They may use different terms to describe the various soil types. However, all soil classification systems aim to provide a comprehensive and consistent way of organizing and categorizing soils based on their physical and chemical properties.

Understanding soil classification is important for understanding the characteristics and potential uses of different soils and for developing appropriate management practices that consider each soil type's unique properties and needs. This is especially important for agriculture, forestry, and other land use activities, as the health and productivity of the soil are directly related to the success of these activities. Soil classification is, therefore, a **vital tool for understanding and managing soil resources** and helps to ensure the long-term health and productivity of the soil.

Overall, soil classification is an important field of study within soil science and helps to understand better and manage soil resources. By understanding the physical, chemical, and biological properties of soils and how these properties interact, soil scientists can develop appropriate management practices that consider the unique characteristics and needs of different soil types. Soil classification is also important for identifying the potential uses of different soils and developing strategies for conserving and protecting soil resources. As soil is a critical natural resource essential for supporting plant growth and maintaining the health of ecosystems, the ongoing study and understanding of soil classification are vital for ensuring our soil resources' long-term health and productivity.

Ultimately, soil formation and classification are important aspects of soil science, as they help us to understand the characteristics and potential uses of different soils. By studying the processes of soil formation and the factors that influence it and classifying soils based on their physical and chemical properties, soil scientists can develop appropriate management practices and advise on the sustainable use of soil resources. Understanding these concepts is essential for ensuring our soil resources' long-term health and productivity.

Chapter Summary

- Soil formation is the process of creating and developing soil over time. Various physical, chemical, and biological factors influence it.
- Physical factors influencing soil formation include climate, topography, and the actions of water, wind, and other weathering agents.

- Chemical factors that influence soil formation include the presence of minerals and organic matter and the effects of acidity and pH.
- Biological factors influencing soil formation include the actions of living organisms, such as plants, animals, and microorganisms.
- Climate, vegetation, geology, and the actions of living organisms are all factors that can influence soil formation.
- Soil classification is the process of organizing and categorizing soils based on their physical and chemical properties.
- Soil classification is important for understanding different soils' characteristics and potential uses and developing appropriate management practices.
- Soils can be classified based on various properties, including texture, structure, color, pH, and nutrient content.

2

SOIL PHYSICAL PROPERTIES

The physical properties of soil are important characteristics that influence its ability to support plant growth and other land use activities. Some of the key physical properties of soil include texture, structure, density, porosity, and water retention and drainage. In this chapter, we will explore these properties in more detail and discuss their significance in soil science.

Texture and Structure of Soil

Soil texture is determined by the relative proportions of soil sand, silt, and clay particles. Sand particles are the largest and tend to be gritty and rough. Silt particles are smaller than sand and have a smooth, floury texture. Clay particles are the smallest and are very fine and smooth. We can determine the relative proportions of these particles in the soil through soil texture analysis, which involves using a set of standardized sieves to separate the particles by size.

The soil's texture can significantly impact its physical properties and ability to support plant growth. For example, soils with a high proportion of sand tend to be well-draining and have good aeration. Still, they may need better water retention and nutrient-holding capacity. Soils with a high proportion of clay tend to have good water retention and nutrient-holding capacity. Still, they may be poorly drained and prone to compaction. Soils with a high proportion of silt tend to have intermediate properties, with good water retention and drainage, but may be prone to erosion.

Soil structure refers to how the particles in soil are arranged and bonded together. Various factors influence soil structure, including the presence of **organic matter,** the actions of **living organisms,** and the influence of **water and other weathering agents.** Soils with a good structure tend to have well-defined pore spaces, which allow for good water and air movement and are more resistant to erosion and compaction. Conversely, soils with a poor structure tend to have poorly defined pore spaces, which can lead to problems with drainage and aeration and may be more prone to erosion and compaction.

Understanding the texture and structure of the soil is important for understanding its physical properties and potential uses. It can also help inform management practices, such as selecting appropriate crops and using appropriate tillage and irrigation techniques. Overall, the texture and structure of soil are important factors that influence its ability to support plant growth and maintain the health of ecosystems.

Density and Porosity of Soil

The soil density is an important factor to consider when studying soil, as it can influence the soil's ability to support plant growth and the movement of water and nutrients through the soil. For example, soil that is too dense may have poor drainage and may be poorly aerated, which can limit the growth of plants. On the other hand, soil that is too porous may not be able to effectively retain water and nutrients, which can also negatively impact plant growth. Soil density and porosity can be affected by various factors, including the type and size of soil particles, the presence of organic matter, and the actions of living organisms.

Understanding soil density and porosity is important for managing soil resources, as it can help to optimize soil conditions for plant growth and to conserve water and nutrients. Techniques such as **tillage**, which involves mechanically breaking up the soil structure, can improve soil density and porosity. Soil amendments, such as adding organic matter or adjusting the pH, can also improve soil density and porosity. In addition, understanding soil density and porosity can be useful for predicting soil behavior under different conditions, such as during periods of drought or heavy rainfall.

Overall, soil density and porosity are important factors to consider when studying soil, as they can influence the soil's ability to support plant growth and retain water and nutrients. By understanding these properties and how we can manage them, soil scientists can help to optimize soil conditions for plant growth and to conserve important resources.

Water Retention and Drainage in Soil

Various factors influence soil water retention and drainage, including soil texture, structure, density, and porosity. For example, soils with a high sand content tend to have poor water retention but good drainage. On the other hand, soils with a high clay content tend to have good water retention but poor drainage. **Soil structure** can also influence water retention and drainage, as soils with a well-developed structure tend to have better water retention and drainage than soils with a poorly developed structure. Finally, soil density and porosity can also influence these properties, as soils with a higher density and lower porosity tend to have poorer water retention and drainage compared to soils with a lower density and higher porosity.

In addition to these physical factors, chemical and biological factors can also influence soil water retention and drainage. For example, the presence of organic matter can help to improve water retention and drainage, as it helps to improve soil structure and increase the number of pores in the soil. In addition, the actions of living organisms, such as earthworms and microorganisms, can also influence water retention and drainage, as they help to break down organic matter and create pores in the soil.

Overall, water retention and drainage are important properties of soil that can significantly impact plant growth and ecosystem health. Therefore, understanding these properties and the factors that influence them is important for managing soil resources and optimizing plant growth.

Ultimately, the physical properties of soil are important characteristics that influence its ability to support plant growth and other land use activities. Understanding these properties and how they can be influenced by factors such as texture, structure, density, porosity, and water retention and drainage is essential for sustainably managing soil resources. By studying the physical properties of soil, soil scientists can develop techniques for conserving and improving soil quality and advise on the appropriate use of soil resources.

Chapter Summary

- The physical properties of soil include texture, structure, density, porosity, and water retention and drainage.
- Soil texture is determined by the relative proportions of soil sand, silt, and clay particles, which we can analyze through soil texture analysis.
- Soil structure refers to how the particles in soil are arranged and bonded together and is influenced by factors such as organic matter, living organisms, and weathering agents.
- Soil density and porosity can influence the soil's ability to support plant growth and the movement of water and nutrients through the soil.
- We can use techniques such as tillage and soil amendments to improve soil density and porosity.
- Soil water retention and drainage are important factors that influence plant growth and ecosystem health.

- Soil water retention is influenced by factors such as soil texture, structure, density, and porosity, as well as the presence of organic matter and soil pH.
- Understanding soil water retention and drainage is important for managing soil resources and predicting soil behavior under different conditions.

3

SOIL CHEMICAL PROPERTIES

T he chemical properties of soil are important characteristics that influence its ability to support plant growth and other land use activities. Some of the key chemical properties of soil include pH, acidity, nutrient content, and soil organic matter. In this chapter, we will explore these properties in more detail and discuss their significance in soil science.

pH and Acidity of Soil

Soil pH is an important factor in soil science, as it can significantly affect plants' health and productivity. Various factors, including the type and amount of minerals present, the presence of organic matter, and the actions of living organisms, can influence the pH of the soil. Some plants are more tolerant of acidic or basic soil conditions than others and may be more suited to certain pH ranges. For example, blueberries and rhododendrons grow best in acidic soil, while grasses and most vegetables prefer a more neutral pH.

Soil pH can be measured using various methods, including pH test strips, pH meters, and laboratory analyses. In addition, we can adjust soil pH by using lime or sulfur, which we can add to the soil to increase or decrease its pH. It is important to maintain an **appropriate pH range** for the plants being grown, as plants may be unable to absorb certain nutrients if the soil pH is too extreme.

Understanding soil pH and its effects on plant growth is an important aspect of soil science and helps to optimize soil fertility and plant productivity. It is also important to understand the ecological relationships between different plants and their soil environments and to develop sustainable land use practices.

Nutrient Content of Soil

Soil nutrient content is an important factor in soil fertility and plant growth. Nutrients such as nitrogen, phosphorus, and potassium are essential for plant growth and development. They are typically classified as macronutrients because they are required in relatively large quantities. Other nutrients, such as micronutrients like zinc and iron, are also important but are needed in much smaller quantities. Soil nutrient content is influenced by various factors, including the type and composition of minerals in the soil, the presence of organic matter, and the soil's pH.

The availability of nutrients to plants is also influenced by soil pH, as certain nutrients are more readily available at certain pH levels. For example, nitrogen and phosphorus are typically more available to plants in soils with a pH between 6 and 7. At the same time, potassium is more readily available in soils with a pH between 5.5 and 7.5. In addition, soil organic matter, composed of decomposed plant and animal material, can also influence nutrient availability by serving as a source of nutrients and improving soil structure and water-holding capacity.

Soil testing is an important tool for assessing soil nutrient content and determining the fertilization needs of plants. Soil test results can help identify any nutrient deficiencies or excesses in the soil and inform the selection of appropriate fertilizers or other soil amendments to address these issues. Proper nutrient management is essential for optimizing plant growth and soil health. In addition, it can help to ensure that soil resources are used sustainably and efficiently.

Soil Organic Matter and Its Role in Soil Fertility

Soil organic matter is formed through decomposition, in which microorganisms and other decomposers break down organic materials. This process releases nutrients that are then made available to plants and help improve the soil's structure and water-holding capacity. Soil organic matter is an important component of soil fertility, as it can help to improve the soil's ability to support plant growth and improve crop yields.

In addition to contributing essential nutrients, soil organic matter also plays a vital role in maintaining the health of the soil ecosystem. It supports the growth of beneficial microorganisms, such as bacteria and fungi, which are essential for breaking down organic matter and releasing nutrients. These microorganisms also help suppress the growth of harmful pathogens and pests, damaging plants and reducing crop yields.

The soil organic matter levels can vary widely depending on factors such as climate, vegetation, and land use practices. Soils rich in organic matter tend to be more fertile and productive and are better able to support plant growth. However, we can lose soil organic matter through various processes, including erosion, leaching, and the removal of organic materials through harvesting and other land use practices. Therefore, it is important to manage soil organic matter sustainably to maintain soil fertility and ecosystem health.

Overall, the chemical properties of soil are important characteristics that influence its ability to support plant growth and other land use activities. Understanding these properties and how they can be influenced by factors such as pH, acidity, nutrient content, and soil organic matter is essential for sustainably managing soil resources. By studying the chemical properties of soil, soil scientists can develop techniques for conserving and improving soil quality and advise on the appropriate use of soil resources.

Chapter Summary

- The chemical properties of soil include pH, acidity, nutrient content, and soil organic matter.
- Soil pH is an important factor influencing plant growth and productivity and can be measured and adjusted using various methods.
- Soil nutrient content is important for plant growth and development. It is influenced by factors such as the type and composition of minerals in the soil, the presence of organic matter, and the soil's pH.
- Soil testing is useful for assessing soil nutrient content and determining fertilization needs.
- Soil organic matter is formed through decomposition and is an important component of soil fertility. As a result, it can improve the soil's ability to support plant growth and crop yields.
- Soil organic matter also plays a vital role in maintaining the soil ecosystem's health and sequestering carbon from the atmosphere.
- Soil organic matter levels can be influenced by various factors, including the type and amount of organic materials added to the soil, the presence of living organisms, and the soil's pH and temperature.
- Understanding and managing soil organic matter are important for optimizing soil fertility and maintaining the soil ecosystem's health.

4

SOIL BIOLOGICAL PROPERTIES

T he biological properties of soil are important characteristics that influence its ability to support plant growth and other land use activities. Some of the key biological properties of soil include the presence of microorganisms and fauna and the role of soil biology in soil health and fertility. In this chapter, we will explore these properties in more detail and discuss their significance in soil science.

Microorganisms in Soil

Microorganisms are essential to the **health and fertility of the soil,** as they help to break down and decompose organic matter, enriching the soil with essential nutrients. They also help to improve soil structure by producing substances that help to bind soil particles together. In addition, some microorganisms can fix nitrogen, converting atmospheric nitrogen into a form that plants use. This process is vital for the growth and productivity of plants, as nitrogen is an essential element required for synthesizing proteins and other important compounds.

Various factors, including soil pH, temperature, and moisture levels, influence the diversity of microorganisms in the soil. Soil management practices, such as using fertilizers, pesticides, and other chemicals, can also impact the populations of microorganisms in the soil. Understanding the role of microorganisms in soil is important for sustainably managing soil resources, as it can help to promote the growth and productivity of plants while also maintaining the health and fertility of the soil.

Fauna in Soil

The presence of microorganisms and fauna in the soil is crucial for maintaining soil health and fertility. These organisms play a vital role in decomposing organic matter and releasing nutrients into the soil, which is essential for plant growth and development. In addition, microorganisms and fauna help create and maintain soil structure, which can improve water retention and drainage and facilitate air and water movement through the soil. These organisms' activity can also help suppress plant diseases and pests, further improving crop yields.

Furthermore, a diverse and healthy community of microorganisms and fauna in the soil can improve soil structure and stability, leading to improved erosion control and reduced soil degradation. This is especially important in **agricultural and forestry settings**, where soil degradation can significantly negatively impact crop yields and ecosystem health. By understanding the role of soil biology in soil health and fertility and implementing management practices that support the growth and activity of these organisms, it is possible to improve the productivity and sustainability of soil resources.

The Role of Soil Biology in Soil Health and Fertility

The health and fertility of the soil are largely determined by the balance and diversity of the biological community within it. When soil contains a diverse and healthy population of microorganisms and fauna, it is more resistant to degradation. As a result, it can support a wider range of plant species. On the other hand, when the biological community in the soil is imbalanced or depleted, it can lead to reduced fertility and increased susceptibility to erosion, compaction, and other forms of degradation.

Soil biology plays a vital role in nutrient cycling, which is essential for maintaining soil fertility. Microorganisms and soil fauna help break down organic matter and release nutrients, such as nitrogen, phosphorus, and potassium, which are essential for plant growth. They also help improve soil structure, which can enhance water retention and improve water infiltration and nutrients into the soil.

In addition to their role in nutrient cycling, soil microorganisms and fauna can also help suppress plant diseases and pests. For example, many soil-dwelling organisms produce antimicrobial compounds that can help control plant disease spread. In addition, some species of insects and worms feed on plant pests, helping to keep their populations in check.

Overall, the health and fertility of the soil are greatly influenced by the biological community within it. Therefore, understanding and managing soil's biological properties make it possible to enhance soil health and fertility and optimize the growth and productivity of crops and other plants.

Ultimately, the biological properties of soil are important characteristics that influence its ability to support plant growth and other land use activities. Understanding these properties and the role of soil biology in soil health and fertility is essential for sustainably managing soil resources. By studying the biological properties of soil, soil scientists can develop techniques for conserving and improving soil quality and advise on the appropriate use of soil resources.

Chapter Summary

- The biological properties of soil include the presence of microorganisms and fauna and the role of soil biology in soil health and fertility.
- Microorganisms in soil are essential for breaking down and decomposing organic matter, enriching the soil with nutrients, and improving soil structure.

- The presence of fauna in the soil is important for decomposing organic matter, releasing nutrients, and maintaining soil structure.
- Soil biology is vital in maintaining soil health and fertility, including nutrient cycling, improving soil structure, and suppressing plant diseases and pests.
- The balance and diversity of the biological community in the soil are important for maintaining soil health and fertility, as an imbalanced or depleted population can lead to soil degradation.
- Soil management practices, such as using fertilizers and pesticides, can impact the populations of microorganisms and fauna in soil.
- Understanding the role of soil biology in soil health and fertility is important for sustainably managing soil resources and improving productivity.
- Techniques such as soil conservation, organic farming, and agroforestry can help promote the soil's health and fertility by supporting the growth and activity of the soil's biological community.

5

SOIL FERTILITY AND NUTRIENT
MANAGEMENT

S oil fertility refers to the ability of soil to support plant growth and
other land-use activities. Various factors, including the availability of nutrients, the presence of organic matter, and the pH
and acidity of the soil, influence soil fertility. Nutrient management is the
process of ensuring that soils have an adequate supply of the nutrients
that are necessary for plant growth. In this chapter, we will explore the
role of nutrients in soil fertility and plant growth, the factors that influence nutrient availability in soil, and techniques for managing soil fertility, including fertilization and soil amendments.

The Role of Nutrients in Soil Fertility and Plant Growth

Various factors, including soil pH, soil organic matter content, and the
presence of certain minerals, influence the availability of nutrients in the
soil. Acidic soils, for example, may have lower levels of available phosphorus. In comparison, alkaline soils may have lower levels of available
iron. Soil organic matter content can also influence nutrient availability,
as it helps release nutrients and act as a buffer to prevent nutrient loss. In
addition, certain minerals in soil may interact with nutrients and influence their availability to plants.

Proper management of soil nutrients is essential for maintaining soil fertility and supporting healthy plant growth. This can include adding **fertilizers** to the soil to supplement nutrient levels and using techniques such as crop rotation and cover cropping to help maintain nutrient levels over time. It is also important to monitor soil nutrient levels and **address any imbalances,** as nutrient deficiencies or excesses can negatively impact plant growth and soil health.

Overall, understanding the role of nutrients in soil fertility and plant growth is important for proper soil management and maintaining healthy and productive soil resources. Understanding the factors that influence nutrient availability and implementing appropriate management practices makes it possible to maintain soil fertility and support healthy plant growth.

Factors that Influence Nutrient Availability in Soil

The presence of certain minerals in the soil can also influence nutrient availability. For example, clay minerals can help hold onto certain nutrients, making them more available to plants. On the other hand, certain minerals, such as aluminum or manganese, can interfere with the availability of other nutrients, such as phosphorus.

The type and amount of vegetation in an area can also influence nutrient availability. For example, plant roots help extract nutrients from the soil, and the type of plants present can influence the types and amounts of nutrients taken up. In addition, the decomposition of plant material can release nutrients back into the soil, making them available to other plants.

Managing soil nutrients is important for maintaining soil fertility and supporting plant growth. This can involve adding nutrient-rich amendments, such as compost or fertilizers, to the soil and practicing sustainable land use practices that help preserve soil nutrients. Understanding the factors influencing nutrient availability in soil is essential for developing effective strategies for managing soil fertility.

Techniques for Managing Soil Fertility

We can use several techniques to manage soil fertility and ensure that soils have an adequate supply of nutrients. These techniques include **fertilization** and the use of **soil amendments**. Fertilization is adding nutrients to the soil as fertilizers, synthetic or organic. Synthetic fertilizers are typically made from chemical compounds. They are designed to provide a specific set of nutrients to the soil. On the other hand, organic fertilizers are made from natural materials such as compost, animal manure, or green manure. They are typically more slowly released into the soil.

Soil amendments are added to soil to improve its physical and chemical properties, such as soil structure, water retention, and nutrient availability. Some common soil amendments include compost, lime, and gypsum. Compost is made from decomposed organic matter, such as plant debris and food waste. It can help to improve soil structure, water retention, and nutrient availability. Lime is a soil amendment made from ground limestone and used to adjust the pH of soil. Gypsum is a soil amendment made from calcium sulfate and is used to improve soil structure and drainage.

In addition to fertilization and the use of soil amendments, other techniques for managing soil fertility include crop rotation and the use of cover crops. Crop rotation is the practice of planting different crops yearly to help improve soil fertility and reduce the risk of soil-borne diseases. Cover crops are crops that are grown specifically to improve soil health rather than for harvest. In addition, cover crops can help to improve soil structure, suppress weeds, and reduce erosion.

Im summary, soil fertility and nutrient management are important aspects of soil science, as they are essential for maintaining the productivity and sustainability of soil resources. By understanding the role of nutrients in soil fertility and plant growth and studying the factors that influence nutrient availability in soil, soil scientists can develop techniques for managing soil fertility and optimizing crop yields. These techniques, including fertilization and soil amendments, can help ensure that soils have an adequate supply of the nutrients necessary for plant growth and development.

Chapter Summary

- Soil fertility refers to the ability of soil to support plant growth and other land-use activities.
- Nutrient management ensures that soils have an adequate supply of the nutrients necessary for plant growth.
- Various factors, including soil pH, soil organic matter content, and certain minerals, influence the availability of nutrients in the soil.
- Proper management of soil nutrients is essential for maintaining soil fertility and supporting healthy plant growth.
- Techniques for managing soil fertility include fertilization and the use of soil amendments.
- Fertilization involves adding nutrients to the soil as fertilizers, synthetic or organic.
- Soil amendments are added to improve the physical and chemical properties of the soil, such as soil structure, water retention, and nutrient availability.
- Other techniques for managing soil fertility include crop rotation and cover crops.

6

SOIL EROSION AND CONSERVATION

S oil erosion is the process by which soil is worn away or removed from an area, often due to the actions of water, wind, or other weathering agents. Soil erosion can significantly impact the productivity and sustainability of soil resources, leading to the loss of fertile soil, decreased water retention, and increased erosion of other natural resources. Soil conservation is protecting soil resources from erosion and other forms of degradation and maintaining or improving soil quality. In this chapter, we will explore the causes and impacts of soil erosion, techniques for preventing and controlling soil erosion, and the role of conservation practices in protecting soil resources.

The Causes and Impacts of Soil Erosion

Soil erosion is the process by which soil is removed from its original location and transported elsewhere, often due to natural processes such as water, wind, and other weathering agents. However, human activities such as poor land management practices, deforestation, and overgrazing can also contribute to soil erosion.

Soil erosion can have serious consequences for the health and productivity of soil and the environment as a whole. It can lead to the loss of fertile soil, which is essential for supporting plant growth and maintaining the health of ecosystems. In addition, water or wind often carries away eroded soil, leading to the deterioration of other natural resources, such as water and air quality. Soil erosion can also contribute to biodiversity loss, as it can alter the habitats of plants and animals.

In addition to these environmental impacts, soil erosion can also have economic consequences, reducing land productivity and making it more difficult to grow crops. It can also increase the risk of natural disasters, such as landslides, which can cause damage to infrastructure and disrupt economic activity.

Overall, soil erosion is a serious problem that requires careful management and conservation efforts to mitigate its negative impacts. This may involve implementing measures such as terracing, contour plowing, and using cover crops to reduce erosion, as well as the restoration of degraded land.

Techniques for Preventing and Controlling Soil Erosion

Preventing and controlling soil erosion is an important aspect of soil management, as it helps to protect and preserve valuable soil resources. One common technique for preventing soil erosion is using physical barriers, such as terracing and windbreaks. Terracing involves the construction of raised beds or ridges on slopes to help prevent soil erosion. In contrast, windbreaks involve planting trees or other vegetation to provide a physical barrier against the wind.

Another technique for preventing soil erosion is using vegetation to stabilize the soil. We can do this by using cover crops, which are planted specifically to protect soil from erosion and improve its overall health. Cover crops can also help to improve water retention and increase soil organic matter.

Mulches are another effective technique for preventing soil erosion. Mulches are materials, such as wood chips or straw, that are spread over the surface of the soil to protect it from erosion and improve water retention. Mulches can also help to regulate soil temperature and suppress weeds, making them an important tool for soil management.

Overall, physical barriers, vegetation, and mulches can effectively prevent and control soil erosion and are important tools for preserving valuable soil resources.

The Role of Conservation Practices in Protecting Soil Resources

The adoption of conservation practices is becoming increasingly important as the global population continues to grow, and the demand for food, fuel, and other natural resources puts pressure on soil resources. Conservation practices help to preserve soil resources for future generations. They can also help mitigate climate change's impacts, as healthy soils are more resistant to drought and other extreme weather events. Some common conservation practices that we can use to protect soil resources include:

- **Crop rotation:** This involves planting different crops in a specific sequence, which can help to prevent soil erosion and depletion of nutrients.
- **Cover cropping:** This involves planting cover crops, such as legumes or grasses, between rows of main crops. Cover crops can help to protect soil from erosion, improve soil structure, and fix nitrogen in the soil.
- **Soil conservation tillage:** This involves using tillage techniques that minimize soil disturbance and reduce the impact of human activities on soil resources. These techniques can include no-till farming, strip tillage, and reduced tillage.
- **Erosion control:** Various techniques can be used to control erosion, such as physical barriers, vegetation, and cover crops. These techniques can help to reduce the impact of water, wind, and other weathering agents on soil resources.
- **Sustainable land management:** This involves adopting practices that balance the needs of humans with the health and sustainability of soil resources. This can include the use of organic farming techniques, the incorporation of conservation

practices into land management plans, and the promotion of sustainable land use practices.

Overall, the adoption of conservation practices is an important step in protecting soil resources and ensuring the long-term sustainability of our natural resources.

Ultimately, soil erosion and conservation are important aspects of soil science, as they protect and preserve soil resources. By understanding the causes and impacts of soil erosion and studying techniques for preventing and controlling soil erosion, soil scientists can develop strategies for conserving and protecting soil resources. The use of conservation practices, such as sustainable land management practices, can also play a critical role in protecting soil resources and ensuring their long-term productivity and sustainability. By promoting the use of erosion control techniques and sustainable land management practices, soil scientists can help protect soil resources from erosion and other degradation forms and maintain or improve soil quality. These efforts are critical for ensuring the long-term productivity and sustainability of soil resources and supporting the growth of plants and other land use activities.

Chapter Summary

- Soil erosion is the process by which soil is removed from its original location and transported elsewhere.
- Human activities, such as poor land management practices, deforestation, and overgrazing, can contribute to soil erosion.
- Soil erosion can lead to the loss of fertile soil and other natural resources, biodiversity loss, and economic consequences.
- Techniques for preventing soil erosion include physical barriers, terracing and windbreaks, vegetation, and mulches.
- Conservation practices, such as crop rotation, cover cropping, soil conservation tillage, erosion control, and sustainable land management, can help to protect soil resources.

- These efforts are critical for ensuring soil resources' long-term productivity and sustainability.
- Conservation practices can help to mitigate climate change's impacts, as healthy soils are more resistant to drought and other extreme weather events.
- By promoting erosion control techniques and sustainable land management practices, soil scientists can help protect soil resources from erosion and other degradation forms.

7

SOIL MANAGEMENT IN AGRICULTURE

S oil management is an important aspect of agriculture, as it is essential for ensuring soil resources' long-term productivity and sustainability. Soil management practices can help to maintain or improve soil quality, optimize crop yields, and promote the growth of plants and other land use activities. In this chapter, we will explore the importance of soil management for sustainable agriculture, techniques for managing soil in different agricultural systems, and the role of soil science in optimizing crop yields and soil health.

The Importance of Soil Management for Sustainable Agriculture

Effective soil management is essential for the long-term sustainability of agricultural systems. Soil is a vital resource that provides the foundation for plant growth and plays a key role in the health and productivity of agricultural systems. However, the soil is a finite resource and is vulnerable to degradation and erosion if not properly managed. Poor soil management practices, such as the overuse of chemical fertilizers and pesticides, can lead to soil degradation, which can decrease soil productivity and negatively impact the environment.

On the other hand, sustainable soil management practices can help maintain or improve soil quality, optimize crop yields, and promote the growth of plants and other land use activities in an environmentally sustainable way. These practices can include using cover crops, mulches, and other soil conservation techniques to reduce erosion and improve soil structure, as well as using organic fertilizers and other sustainable land management practices.

Effective soil management is also essential for addressing global challenges such as food security and climate change. Soil management practices that enhance soil fertility and promote sustainable agriculture can help to increase crop yields and improve food security in a changing climate. In addition, soil management practices that promote carbon sequestration and reduce greenhouse gas emissions can help to mitigate climate change and contribute to the long-term sustainability of agricultural systems. Overall, soil management is a critical component of sustainable agriculture and is essential for ensuring our agricultural systems' long-term health and productivity.

Techniques for Managing Soil in Different Agricultural Systems

There are a variety of techniques that we can use to manage soil in different agricultural systems. These techniques can vary depending on the specific needs and characteristics of the soil and the type of crops being grown. Some common soil management practices in agriculture include using cover crops, mulches, and fertilizers, as well as implementing sustainable land management practices such as conservation tillage and integrated pest management. These practices can help to improve soil structure, fertility, and productivity, as well as reduce the impact of agriculture on the environment.

One technique that is commonly used to manage soil in agriculture is the use of **cover crops.** Cover crops are grown specifically to protect soil and improve its physical and chemical properties. For example, cover crops can help to prevent erosion, reduce the need for chemical fertilizers, and improve soil structure and water retention. They can also help suppress weeds and pests and provide a habitat for beneficial insects and other wildlife.

Mulches are another important tool for soil management in agriculture. Mulches are materials, such as straw or wood chips, applied to the soil surface to help protect it from erosion, retain moisture, and suppress weeds. Mulches can be used in various agricultural systems, including row crops, orchards, and gardens.

Fertilizers are another common soil management practice in agriculture. Fertilizers are materials applied to soil to give plants the nutrients they need to grow and develop. Many different types of fertilizers are available, including synthetic fertilizers and organic fertilizers. Synthetic fertilizers are made from inorganic chemicals, while organic fertilizers are made from natural materials such as compost or animal manure.

Sustainable land management practices, such as conservation tillage and integrated pest management, are also important tools for managing soil in agriculture. Conservation tillage is a technique that involves minimizing soil disturbance during planting and harvesting, which can help to reduce erosion and improve soil structure. Integrated pest management is a technique that involves using a variety of strategies, such as biological control and cultural practices, to manage pests in a way that is sustainable and environmentally friendly.

We can use many techniques to manage soil in different agricultural systems. The most appropriate techniques will depend on the specific needs and characteristics of the soil and the crops being grown. Using a combination of cover crops, mulches, fertilizers, and sustainable land management practices, we can optimize soil productivity and reduce the impact of agriculture on the environment.

The Role of Soil Science in Optimizing Crop Yields and Soil Health

Soil science is critical in optimizing crop yields and soil health in agricultural systems. By studying soil's physical, chemical, and biological properties, soil scientists can develop techniques for managing soil in a way that maximizes crop yields and promotes soil health. This can involve using fertilizers, soil amendments, and other techniques to optimize nutrient availability, improve soil structure, and promote the growth of beneficial microorganisms.

Im summary, soil management is an essential aspect of agriculture, as it is critical for ensuring the long-term productivity and sustainability of soil resources. By understanding the importance of soil management for sustainable agriculture and studying techniques for managing soil in different agricultural systems, soil scientists can develop strategies for optimizing crop yields and promoting soil health. The role of soil science in this process is critical, as it provides the knowledge and tools needed to understand and manage soil resources in a productive and sustainable way. Furthermore, by applying this knowledge, soil scientists can help to ensure that agricultural systems can support the growth of plants and other land use activities in a sustainable and environmentally responsible way.

Chapter Summary

- Soil management is essential for the long-term sustainability of agricultural systems.
- Poor soil management practices can lead to soil degradation, negatively affecting the environment.
- Sustainable soil management practices can help to maintain or improve soil quality and optimize crop yields.
- Effective soil management is also essential for addressing global challenges such as food security and climate change.
- Techniques for managing soil in different agricultural systems include using cover crops, mulches, and fertilizers and implementing sustainable land management practices.
- Soil science is critical in optimizing crop yields and soil health in agricultural systems.
- By studying soil's physical, chemical, and biological properties, soil scientists can develop techniques for managing soil to maximize crop yields and promote soil health.
- Soil science provides the knowledge and tools to understand and manage soil resources productively and sustainably.

8

SOIL MANAGEMENT IN FORESTRY
AND LANDSCAPING

S oil management is an important aspect of forestry and landscaping, as it is essential for ensuring the long-term productivity and sustainability of soil resources in these settings. In addition, soil management practices can help to maintain or improve soil quality, optimize the growth of trees and other plants, and promote the health and beauty of landscapes. In this chapter, we will explore the role of soil science in managing forest and landscape soils, techniques for conserving and improving soil quality in these settings, and the impact of land use practices on soil resources.

The Role of Soil Science in Managing Forest and Landscape Soils

Soil science plays a critical role in managing forest and landscape soils, as it provides the knowledge and tools needed to understand and manage soil resources in these settings. By studying soil's physical, chemical, and biological properties, soil scientists can develop techniques for managing soil to maximize the growth of trees and other plants and promote soil health. This can involve the use of fertilizers, soil amendments, and other techniques to optimize nutrient availability, improve soil structure, and promote the growth of beneficial microorganisms.

Techniques for Conserving and Improving Soil Quality in these Settings

There are a variety of techniques that we can use to conserve and improve soil quality in forestry and landscaping settings. These techniques can vary depending on the specific needs and characteristics of the soil and the type of plants being grown. Some common soil management practices in these settings include using cover crops, mulches, and fertilizers, as well as implementing sustainable land management practices such as conservation tillage and integrated pest management.

To learn more about these soils management practices, revisit chapter 7.

The Impact of Land Use Practices on Soil Resources

How land is used can have a significant impact on soil resources. For example, poor land use practices, such as deforestation and overgrazing, can lead to soil degradation, decreasing soil productivity and negatively impacting the environment. By contrast, sustainable land use practices, such as sustainable forestry and landscaping, can help maintain or improve soil quality, optimize the growth of trees and other plants, and promote the health and beauty of landscapes in a sustainable way.

In summary, soil management is an essential aspect of forestry and landscaping, as it is critical for ensuring the long-term productivity and sustainability of soil resources in these settings. By understanding the role of soil science in managing forest and landscape soils and studying techniques for conserving and improving soil quality in these settings, soil scientists can develop strategies for optimizing the growth of trees and other plants and promoting soil health. The impact of land use practices on soil resources is also an important consideration, as how land is used can have significant consequences for soil quality and productivity. By applying this knowledge, soil scientists can help ensure that forestry and landscaping practices can support the growth of trees and other plants in a sustainable and environmentally responsible way.

Chapter Summary

- Soil science plays a critical role in managing forest and landscape soils.
- There are a variety of techniques that we can use to conserve and improve soil quality in forestry and landscaping settings.
- Poor land use practices can lead to soil degradation, while sustainable land use practices can help maintain or improve soil quality.
- Soil management is essential for ensuring the long-term productivity and sustainability of soil resources in these settings.
- By understanding the role of soil science and studying techniques for conserving and improving soil quality, soil scientists can develop strategies for optimizing the growth of trees and other plants and promoting soil health.
- The impact of land use practices on soil resources is also an important consideration.
- By applying this knowledge, soil scientists can help ensure that forestry and landscaping practices can support the growth of trees and other plants in a sustainable and environmentally responsible way.

9

SOIL MANAGEMENT IN ENVIRONMENTAL RESTORATION

Environmental restoration is rehabilitating degraded or damaged ecosystems to return them to a more natural and functional state. Soil management is an important aspect of environmental restoration. The soil is a vital component of ecosystems. It is critical for supporting the growth of plants and other land-use activities. In this chapter, we will explore the role of soil science in environmental restoration projects, techniques for rehabilitating degraded soils, and the importance of soil health in maintaining ecosystem function.

The Role of Soil Science in Environmental Restoration Projects

Soil science is a key discipline in environmental restoration, as it helps to understand and manage soil resources to promote the growth of plants and restore ecosystem function. Soil scientists use various techniques to improve soil quality, including using fertilizers, soil amendments, and sustainable land management practices such as conservation tillage and integrated pest management.

These techniques are designed to **optimize nutrient availability, improve soil structure,** and **promote the growth of beneficial microorganisms,** all of which are critical for the success of environmental restoration projects. In addition, soil scientists also work to understand the impacts of environmental stressors, such as pollution and climate change, on soil health to develop strategies for mitigating these impacts and promoting soil recovery. Overall, the role of soil science in environmental restoration is crucial in ensuring the long-term sustainability of our natural resources.

Techniques for Rehabilitating Degraded Soils

To effectively rehabilitate degraded soils, assessing the underlying causes of soil degradation is important. This can involve studying the soil's physical, chemical, and biological properties, as well as evaluating the potential impact of human activities such as deforestation and overgrazing. Once the causes of soil degradation have been identified, appropriate techniques can be selected and implemented to restore soil quality and promote the growth of plants and other land-use activities.

One important aspect of soil rehabilitation is the use of cover crops and mulches. These techniques involve planting crops or applying organic materials to the soil surface, which can help improve soil structure, reduce erosion, and promote the growth of beneficial microorganisms. Cover crops and mulches can also help to improve soil water retention, which is important for supporting plant growth in areas with limited water resources.

In addition, to cover crops and mulches, fertilizers and other soil amendments can be used to improve soil nutrient availability and promote plant growth. However, it is important to use these techniques judiciously, as **over-fertilization** can lead to soil degradation and negative environmental impacts. Instead, it is often more effective to implement sustainable land management practices that take into account the long-term health and productivity of the soil. This can include techniques such as conservation tillage, which involves minimizing the disturbance of soil and preserving soil structure, and integrated pest management, which involves using natural methods for controlling pests and diseases.

The Importance of Soil Health in Maintaining Ecosystem Function

Soil health is critical for maintaining ecosystem function, as it is essential for supporting the growth of plants and other land-use activities. Degraded soils can *negatively* impact ecosystem function, as they may be less able to support the growth of plants and other land use activities. By rehabilitating degraded soils and promoting soil health, it is possible to sustainably restore ecosystem function and support the growth of plants and other land use activities.

In summary, soil management is an important aspect of environmental restoration, as it is critical for ensuring the long-term productivity and sustainability of soil resources in these settings. By understanding the role of soil science in environmental restoration projects and studying techniques for rehabilitating degraded soils, soil scientists can develop strategies for restoring ecosystem function and promoting the growth of plants and other land use activities. The importance of soil health in maintaining ecosystem function is also an important consideration, as soil health is essential for sustainably supporting the growth of plants and other land-use activities. By applying this knowledge, soil scientists can help to ensure that environmental restoration efforts can restore ecosystems practically and sustainably.

Chapter Summary

- Environmental restoration is rehabilitating degraded or damaged ecosystems to return them to a more natural and functional state.
- Soil science is a key discipline in environmental restoration, as it helps to understand and manage soil resources to promote the growth of plants and restore ecosystem function.
- Techniques for rehabilitating degraded soils include using fertilizers, soil amendments, and sustainable land management practices such as conservation tillage and integrated pest management.

- Cover crops and mulches are important techniques for improving soil structure, reducing erosion, and promoting the growth of beneficial microorganisms.
- Fertilizers and other soil amendments can improve nutrient availability and promote plant growth.
- Soil health is critical for maintaining ecosystem function, as it is essential for supporting the growth of plants and other land-use activities.
- By understanding the role of soil science in environmental restoration projects and studying techniques for rehabilitating degraded soils, soil scientists can develop strategies for restoring ecosystem function and promoting the growth of plants and other land-use activities.
- The importance of soil health in maintaining ecosystem function is essential for sustainably supporting the growth of plants and other land-use activities.

10

SOIL MANAGEMENT IN URBAN AND INDUSTRIAL SETTINGS

S oil management is an important aspect of urban and industrial areas, as it is essential for ensuring the long-term productivity and sustainability of soil resources in these settings. In addition, soil management practices can help to maintain or improve soil quality, optimize the growth of plants and other land use activities, and promote the health and beauty of urban and industrial environments. In this chapter, we will explore the challenges of managing soil in urban and industrial areas, techniques for conserving and improving soil quality in these settings, and the role of soil science in designing and managing green spaces in urban areas.

The Challenges of Managing Soil in Urban and Industrial Areas

Managing soil in urban and industrial areas can be challenging, as these environments often have unique soil conditions and characteristics that can be difficult to manage. In addition, urban and industrial soils may be subjected to various stressors, such as pollution, compaction, and alteration, which can negatively impact soil quality and productivity. For example, we may contaminate urban and industrial soils with heavy metals or other toxic substances, which can make them inhospitable to plants and other living organisms.

Additionally, the lack of space and other constraints in urban and industrial areas can make it difficult to implement traditional soil management practices, such as using cover crops and mulches or applying fertilizers and other soil amendments. These challenges highlight the importance of developing innovative approaches to soil management in urban and industrial areas to ensure that these soils can support the growth of plants and other land-use activities in a sustainable way.

Techniques for Conserving and Improving Soil Quality in these Settings

Managing soil in urban and industrial areas can be challenging due to the unique conditions and characteristics that these environments often present. Urban and industrial soils may be subjected to various stressors such as pollution, compaction, and alteration, which can affect soil quality and productivity negatively. In addition, urban and industrial areas' space constraints and other factors can make it difficult to implement traditional soil management practices.

Therefore, it is important to use techniques specifically tailored to these environments' needs to conserve and improve soil quality. Some common techniques include using cover crops, mulches, and fertilizers, as well as implementing sustainable land management practices such as conservation tillage and integrated pest management. These techniques can help to maintain or improve soil quality, optimize plant growth, and promote sustainable land use in urban and industrial areas.

To learn more about these soils management practices, revisit chapter 7.

The Role of Soil Science in Designing and Managing Green Spaces in Urban Areas

Soil science plays a critical role in designing and managing green spaces in urban areas, as it provides the knowledge and tools needed to understand and manage soil resources in these settings. By studying the physical, chemical, and biological properties of soil, soil scientists are able to develop techniques for managing soil in a way that promotes the growth of plants and other land-use activities and enhances the health and beauty of urban environments. This can involve using fertilizers, soil amendments, and other techniques to optimize nutrient availability, improve soil structure, and promote the growth of beneficial microorganisms.

In summary, soil management is an important aspect of urban and industrial areas, as it is critical for ensuring the long-term productivity and sustainability of soil resources in these settings. By understanding the challenges of managing soil in urban and industrial areas and by studying techniques for conserving and improving soil quality in these settings, soil scientists can develop strategies for optimizing the growth of plants and other land-use activities and promoting the health and beauty of urban environments. The role of soil science in designing and managing green spaces in urban areas is also important, as soil science provides the knowledge and tools needed to understand and manage soil resources in these settings practically and sustainably.

Chapter Summary

- Soil management is an important aspect of urban and industrial areas.
- Managing soil in urban and industrial areas can be challenging due to the unique conditions and characteristics of these environments.
- Techniques such as using cover crops, mulches, and fertilizers, as well as implementing sustainable land management practices, can help to conserve and improve soil quality.
- Soil science plays a critical role in designing and managing green spaces in urban areas, as it provides the knowledge and tools needed to sustainably understand and manage soil resources.

CONCLUSION

S oil science is a critical field that is essential for understanding and managing soil resources. Soil is a vital component of ecosystems and is critical for supporting the growth of plants and other land-use activities. In this final chapter, we will explore the importance of soil science in understanding and managing soil resources, and the ongoing need for research and innovation in the field.

The Importance of Soil Science in Understanding and Managing Soil Resources

Soil science plays a critical role in understanding and managing soil resources, as it provides the knowledge and tools needed to understand and manage soil in a productive and sustainable way. By studying soil's physical, chemical, and biological properties, soil scientists can develop techniques for managing soil in a way that promotes the growth of plants and other land use activities and maintains or improves soil quality. This knowledge is essential for a variety of applications, including agriculture, forestry and landscaping, environmental restoration, and urban and industrial settings.

The Ongoing Need for Research and Innovation in the Field

Despite the importance of soil science, much still needs to be discovered about soil and its role in ecosystems. As a result, there is a continuing need for research and innovation in the field of soil science. By studying the properties of soil and developing new techniques for managing soil resources, soil scientists can improve our understanding of soil and its role in supporting the growth of plants and other land-use activities. This knowledge is critical for developing strategies for optimizing soil productivity and sustainability and addressing the future challenges that soil resources face.

In summary, soil science is a required field that is essential for understanding and managing soil resources. By studying soil's physical, chemical, and biological properties, soil scientists can develop techniques for managing soil in a way that promotes the growth of plants and other land use activities and maintains or improves soil quality. The ongoing need for research and innovation in soil science is also important, as it allows us to continue to improve our understanding of soil and its role in ecosystems and to develop new strategies for optimizing soil productivity and sustainability. By applying this knowledge, soil scientists can make important contributions to a wide range of fields and help ensure that soil resources can support the growth of plants and other land-use activities in a sustainable way.

AFTERWORD

The afterword to this book is a call to action. We have discussed the power of regenerative agriculture and soil science. Still, it is up to us to put this knowledge into practice. The health of our planet is at stake, and we must work together to ensure that it is sustained for future generations.

We must work to promote sustainable farming practices, invest in soil science research, and educate the public about the importance of soil health and regenerative agriculture. By doing this, we can ensure that our planet is healthy and thriving for many years.

Thank you for joining us on this journey to a healthier future. Together, we can create a better world for our children and grand-children.

ACKNOWLEDGMENTS

I want to express my sincerest gratitude to all who have contributed to making this 2-in-1 collection.

First and foremost, I thank my editor and publisher for their patience, guidance, and expertise. Without their help, this book would not be the same.

I would also like to thank the many farmers and researchers who shared their stories and data. Their contributions were invaluable to the research and development of this book.

Finally, I thank my family and friends for their constant support and encouragement. They were my rock throughout this journey; I could not have done it without them.

ABOUT THE AUTHOR

Michael Barton is an expert in sustainability and regenerative agriculture, with over a decade of experience in the field. He holds a degree in Environmental Studies and has received advanced training in the principles and practices of regenerative agriculture.

Michael has worked with farmers, policymakers, and environmentalists to promote sustainable farming practices and advocate for the adoption of regenerative agriculture. In addition, he has collaborated with organizations worldwide to advance sustainable agriculture and food systems.

Michael is the author of the 2-in-1 collection *"Fostering a Healthy Planet,"* which offers a wealth of agriculture and soil science knowledge. The first book, *"The Power of Regenerative Agriculture,"* explores the principles and practices of regenerative agriculture. In contrast, the second book, *"Introduction to Soil Science,"* is an in-depth look into the scientific study of soil.

Michael's passion for sustainability and his belief in the importance of soil health shine through on every page of this valuable resource. He provides a comprehensive guide to the fundamental principles of soil science, focusing on practical applications and equipping readers with the knowledge and skills to effectively understand and manage soil resources.

FROM THE AUTHOR

Dear Reader,

I hope you enjoyed the book! I would love to hear your feedback—I personally read every review. If you have three minutes to help me, can you please post a short review on the platform from which you purchased this book?

Also, I would be honored if you would sign up for my mailing list to keep updated on my upcoming books, exclusive content, and special promotions. Plus, as a subscriber, you'll have the opportunity to **download my new books for free** in exchange for an honest review. Scan to QR code to sign up.

SCAN ME

Best, Michael Barton

www.ingramcontent.com/pod-product-compliance
Lightning Source LLC
Chambersburg PA
CBHW071426210326
41597CB00020B/3668